Application of SERS for Nanomaterials

Application of SERS for Nanomaterials

Editors

Ronald Birke
Bing Zhao

MDPI • Basel • Beijing • Wuhan • Barcelona • Belgrade • Manchester • Tokyo • Cluj • Tianjin

Editors
Ronald Birke
City University of New York (CUNY)
USA

Bing Zhao
Jilin University
China

Editorial Office
MDPI
St. Alban-Anlage 66
4052 Basel, Switzerland

This is a reprint of articles from the Special Issue published online in the open access journal *Nanomaterials* (ISSN 2079-4991) (available at: https://www.mdpi.com/journal/nanomaterials/special_issues/SERS_Nanomaterials).

For citation purposes, cite each article independently as indicated on the article page online and as indicated below:

LastName, A.A.; LastName, B.B.; LastName, C.C. Article Title. *Journal Name* **Year**, *Volume Number*, Page Range.

ISBN 978-3-0365-2792-5 (Hbk)
ISBN 978-3-0365-2793-2 (PDF)

Contents

About the Editors

Ronald L. Birke received degrees from the Chemistry Departments of the University of North Carolina (Chapel Hill) (B.S.) and the Massachusetts Institute of Technology (Ph.D.) and completed his postdoctoral studies at the Free University of Brussels, where his mentors were Charles N. Reilley, David K. Roe, and Lucien Gierst. He was an instructor at Harvard University and an Associate Professor at the University of South Florida before coming to the City College of the City University of New York (CUNY) in 1974 as an Associate Professor. In 1981, he was promoted to full Professor, and he is currently emeritus Professor of Chemistry and a member of the Ph.D. Program in Chemistry at the Graduate School of CUNY. His research interests are computational quantum chemistry studies of molecules on surfaces, especially regarding simulations of Surface-Enhanced Raman spectroscopy.

Bing Zhao received his B.S. degree in 1984 at the Department of Chemistry, Jilin University. Following this, he completed his graduate studies to the Institute of Theoretical Chemistry at Jilin University and obtained his M.S.and Ph.D. degrees in 1987 and 1992, respectively. In 1992, he joined the Chemistry Faculty of the State Key Laboratory of Supramolecular Structure and Materials at Jilin University. In 1998, he was promoted to full professor at the State Key Laboratory of Supramolecular Structure and Materials. His current research concerns vibrational spectroscopy and spectroscopic analysis, especially as applied to the field of SERS spectroscopy.

 *nanomaterials* MDPI

Editorial

Special Issue: Application of SERS for Nanomaterials

Ronald L. Birke

Department of Chemistry and Biochemistry, The City College of the City University of New York, 160 Convent Avenue, New York, NY 10031, USA; rbirke@ccny.cuny.edu

Surface-enhanced Raman scattering (SERS) is now a relatively mature field of spectroscopy, with it having been almost 50 years since its first experimental demonstration [1]. A main feature of SERS is the requirement for a surface substrate that facilitates the enhancement mechanism. This surface-enhancing substrate can produce a plasmon resonance electromagnetic enhancement of the stimulating light and/or facilitate a photon driven charge-transfer (CT) resonance [2,3]. The combined plasmonic and CT resonance process occurs on substrates of metal nanoparticles (NPs) such as Ag, Au, Cu, and other metals [4], whereas a photon driven charge-transfer resonance (CT-SERS) is the main enhancing process on semiconductor nanoparticles [5]. Within the variety of enhancement processes possible, an additional scattering mechanism involves a combination of the plasmonic resonance with a molecular resonance sometimes called surface-enhanced resonance Raman scattering (SERRS) [6]. The molecular resonance generates Franck–Condon scattering while the charge transfer resonance generates Herzberg–Teller scattering [2,3,7]. This diversity of scattering mechanisms makes the formulation of a surface Raman theory, which covers all cases, challenging. The various enhancement mechanisms, all involving engineered or assembled nanoparticle surfaces, can boost the relatively low normal Raman signal to substantial and possibly "giant" intensities, even allowing for single-molecule detection in the most efficient cases [8]. Furthermore, the rich Raman vibrational spectra obtained with SERS methods provide molecular level information and chemical identity. These properties have established SERS methodologies as ultra-sensitive techniques for analytical sensing of chemical and biochemical compounds and for the investigation of the properties of nanostructured plasmonic metal [9] and semiconductor dielectric substrates [10,11]. Since the sensing process is linked to progress in the design and fabrication of new nanomaterial surfaces, a recent Special Issue of this journal was specifically devoted to this aspect of SERS development [12]. However, SERS is still being developed in terms of the fabrication of new nanomaterial surface substrates, new applications in chemical and biochemical sensing, improvement of methods for interrogating surface structure such as imaging with tip-enhanced Raman scattering (TERS), studies of chemical reactions with SERS, and the computational simulation of the surface Raman scattering spectra at different nanomaterial interfaces. The present Special Issue includes research papers which address these areas of SERS developments.

Citation: Birke, R.L. Special Issue: Application of SERS for Nanomaterials. *Nanomaterials* **2021**, *11*, 3300. https://doi.org/10.3390/nano11123300

Received: 26 November 2021
Accepted: 2 December 2021
Published: 6 December 2021

Publisher's Note: MDPI stays neutral with regard to jurisdictional claims in published maps and institutional affiliations.

There are eight research papers in this Special Issue which explore important developments for future applications of SERS. The first paper by Pei Dai et al. [13] illustrates the fabrication of inexpensive hydrophobic pure Cu chips and those modified with a thin shell of Ag. The chips are made by depositing the Cu nanoparticles on a flexible fabric support by chemical reduction methods with the Ag layer formed by a replacement method. The Ag-Cu chip substrates show a higher enhancement factor with good stability, maintaining 80% of its intensity for up to two months. Several organic molecules are investigated and separated into two groups based on their HOMO/LUMO levels with respect to the equilibrated Fermi level of Cu-Ag and the possibility of photoinduced charge transfer, PICT. Additionally, there are three papers involving the charge transfer (CT) mechanism for wide-band gap semiconductors-ZrO_2 [14], TiO_2 [15], and MoO_3 [16]. Yi et al. [14] investigated two crystal forms of ZrO_2 as substrates for SERS. Raman spectral shifts indicate that the scattering mechanism is a CT mechanism. Using hydrothermal synthesis with additives for selective growth, they show that the 99.7% tetragonal phase is superior to the monoclinic phase since it has highest degree of charge transfer parameter. The energy levels of surface defect states in the band gap of ZrO_2 play a decisive role in rationalizing the CT scattering mechanism from ZrO_2 to the LUMO of 4-mercaptobenzoic acid (4–MBA). Birke and Lombardi [15] use Density Functional Theory (DFT) to simulate the surface Raman spectra of the solar cell dye N3 adsorbed on a Ti_5O_{10} nanoparticle cluster model. This small TiO_2 cluster was found to have similar properties to clusters as large as $Ti_{38}O_{76}$. Enhancement factors of about 1×10^3 and 2×10^2 for resonance Raman and charge transfer surface scattering mechanisms, respectively, were found from the simulations. The CT-SERS simulation shows that a direct dye to semiconductor photoinduced electron transfer is possible without going through an intermediate dye excited state, which is relevant to the excitation mechanism of the dye-sensitized solar cell (DSSC). In the paper by Chu et al. [16], a composite system is fabricated by co-sputtered Ag/MoO_3 on polystyrene microspheres which contains abundant oxygen vacancy defects. 4-Aminothiophenol (PATP) is used as a SERS probe molecule and the oxygen vacancy defects in MO_3 are indicated as intermediate energy channels for electron transport between the Fermi level of Ag and the LUMO of PATP. Another study in this issue of a composite system by Lin Guo et al. [17] utilizes 4-mercaptobenzoic acid (MBA) as a bridge between Au nanorods (NRs) of different length to diameter ratios (L/D) and a Cu_2O semiconductor with a consistent 15 nm thickness in a core–shell (Au NR–MBA@Cu_2O) structure. Here, the sulfur end of the molecule binds to the Au atom of the nanorod and the carboxylate end binds to Cu^+ of the Cu_2O shell. The L/D ratios of the NRs adjust the surface plasmon resonance (SPR) and the specific surface area of the nanorods. These assemblies are investigated for plasmon absorption characteristics, changes in SERS bands, and for the degree of charge transfer as a function of L/D ratio. The CT takes place from the Au NRs through the LUMO level of MBA to the conduction band (CB) of the Cu_2O shell. When the wavelength of the SERS incident laser light is close to the maximum of the longitudinal SPR absorption curve, the degree of the charge transfer process increases. Furthermore, increased surface area of Au NRs correlates with the movement of CB of Cu_2O closer to the LUMO level of MBA facilitating the degree of charge transfer. These studies show the value of SERS in elucidating the basic photo-physics in nanomaterial systems.

The method of TERS allows for imaging below the Abbe half-wavelength diffraction limit of the imaging light and can show resolutions in the tens of nanometers. This method has a significant future for revealing new chemistry and physics in nanostructured systems. The paper of Mandelbaum et al. [18] explores the effect of shape (hemisphere, hemispheroid, ellipsoidal cavity, ellipsoidal rod, nano-cone) and material (Ag, Au, Al) on surface enhancement using the finite element numerical and analytical approximation methods for single tip structures, a four-probe configuration, and for pixel arrays of tips. The hemispheroid was recommended as the best geometry for the nanoparticle tips.

The remaining two articles in this Special Issue [19,20] show noteworthy applications of SERS for biochemical and chemical sensing. Zavyalova et al. [19] develop a simple rapid quantitative colloidal Ag based SERS method for the detection of the COVID-19 coronavirus SARS-CoV-2. The selectivity is obtained with the DNA RBD-1C aptamer which selectively binds the receptor binding domain (RBD) of the surface S-protein of the virus, and the detection sensitivity comes from the Bodipy Fl SERS reporter molecule. The essence of this assay is that the SERS intensity increases in the presence of increasing SARS-CoV-2 but decreases with increasing non-specific viral particles. The method is one-step and fast (7 min), with a high sensitivity (LOD of 5.5 Å~10^4 TCID50/mL). In the final article in this Special Issue, Xu et al. [20] studies the SERS detection methodology for the notorious herbicide and crop desiccant 'glyphosate' (GLP) or N-(phosphonomethyl)gycine. This compound is the active ingredient in weed killers such as 'Roundup' and has been claimed to cause non-Hodgkin's lymphoma. An isotopic glyphosate denoted 13-GLP with ^{13}C substitution at the 2-position, 2-^{13}C-glyphosate, as-well-as the non-isotopic 12-GLP are investigated. The method is based on the ninhydrin reaction with the amino group of GLP catalyzed by MoO_4 to form a purple color dye product, PD. The PD products formed from both 12-GLP and 13-GLP are studied. Raman and Fourier transform infrared (FTIR) spectra are obtained in the absence of Ag nanoparticles (NPs) and SERS spectra are obtained in the presence of Ag NPs. DFT calculations at the B3LYP/6-311++G(d,p) level are used to interpret the vibrational band assignments of the spectra. The use of isotopic substitution is recommended as a promising area for future use in the interpretation of SERS spectra.

Funding: This article received no external funding.

Institutional Review Board Statement: Not applicable.

Informed Consent Statement: Not applicable.

Data Availability Statement: Not applicable.

Acknowledgments: We acknowledge all of those who have contributed to this Special Issue especially Zhao Bing for his kind assistance in the organization of this Special Issue.

Conflicts of Interest: The author declares no conflict of interest.

References

1. Fleishman, M.; Hendra, P.J.; MacQuillian, A.J. Raman Spectra of Pyridine Adsorbed at a Silver Electrode. *Chem. Phys. Lett.* **1974**, *26*, 163–166. [CrossRef]
2. Lombardi, J.R.; Birke, R.L.; Lu, T.; Xu, J. Charge-Transfer Theory of Surface Enhanced Raman Spectroscopy: Herzberg-Teller Contributions. *J. Chem. Phys.* **1986**, *84*, 4174–4180. [CrossRef]
3. Lombardi, J.R.; Birke, R.L. A Unified Approach to Surface-Enhanced Raman Spectroscopy. *J. Phys. Chem. C* **2008**, *112*, 5605. [CrossRef]
4. Fernanda Cardinal, M.; Vander Ende, E.; Hackler, R.A.; McAnally, M.O.; Stair, P.C.; Schatz, G.C.; Van Duyne, R.P. Expanding Applications of SERS through Versatile Nanomaterials Engineering. *Chem. Soc. Rev.* **2017**, *46*, 3886. [CrossRef] [PubMed]
5. Lombardi, J.R.; Birke, R.L. Theory of Surface-Enhanced Raman Scattering in Semiconductors. *J. Phys. Chem. C* **2014**, *118*, 11120–11130. [CrossRef]
6. Lombardi, J.R.; Birke, R.L. A Unified View of Surface-Enhanced Raman Scattering. *Acc. Chem. Res.* **2009**, *42*, 734–742. [CrossRef] [PubMed]
7. Birke, R.L.; Lombardi, J.R. Relative Contributions of Franck-Condon to Herzberg-Teller Terms in Charge Transfer Surface-Enhanced Raman Scattering Spectroscopy. *J. Chem. Phys.* **2020**, *152*, 224107. [CrossRef] [PubMed]
8. Kneipp, J.; Kneipp, H.; Kneipp, K. SERS—A Single-Molecule and Nanoscale Tool for Bioanalytics. *Chem. Soc. Rev.* **2008**, *37*, 1052–1060. [CrossRef] [PubMed]
9. Shvalya, V.; Filipič, G.; Zavasnik, J.; Abdulhalim, I.; Cvelbar, U. Surface-Enhanced Raman Spectroscopy for Chemical and Biological Sensing Using Nanoplasmonics: The Relevance of Interparticle Spacing and Surface Morphology. *Appl. Phys. Rev.* **2020**, *7*, 031307. [CrossRef]
10. Han, X.X.; Ji, W.; Zhao, B.; Ozaki, Y. Semiconductor-Enhanced Raman Scattering: Active Nanomaterials and Applications. *Nanoscale* **2017**, *9*, 4847. [CrossRef] [PubMed]
11. Alessandri, I.; Lombardi, J.R. Enhanced Raman Scattering with Dielectrics. *Chem. Rev.* **2016**, *116*, 14921–14981. [CrossRef] [PubMed]
12. Barbillon, G. Application of Novel Plasmonic Nanomaterials on SERS. *Nanomaterials* **2020**, *10*, 2038. [CrossRef] [PubMed]
13. Dai, P.; Li, H.; Huang, X.; Wang, N.; Zhu, L. Highly Sensitive and Stable Copper-Based SERS Chips Prepared by a Chemical Reduction Method. *Nanomaterials* **2021**, *11*, 2770. [CrossRef] [PubMed]
14. Yi, M.; Zhang, Y.; Xu, J.; Deng, D.; Mao, Z.; Meng, X.; Shi, X.; Zhao, B. Surface Enhanced Raman Scattering Activity of ZrO_2 Nanoparticles: Effect of Tetragonal and Monoclinic Phases. *Nanomaterials* **2021**, *11*, 2162. [CrossRef] [PubMed]
15. Birke, R.L.; Lombardi, J.R. DFT and TD-DFT Investigation of a Charge Transfer Surface Resonance Raman Model of N3 Dye Bound to a Small TiO_2 Nanoparticle. *Nanomaterials* **2021**, *11*, 1491. [CrossRef] [PubMed]
16. Chu, Q.; Li, J.; Jin, S.; Guo, S.; Park, E.; Wang, J.; Chen, L.; Jung, Y.M. Charge-Transfer Induced by the Oxygen Vacancy Defects in the Ag/MoO_3 Composite System. *Nanomaterials* **2021**, *11*, 1292. [CrossRef] [PubMed]
17. Guo, L.; Mao, Z.; Jin, S.; Zhu, L.; Zhao, J.; Zhao, B.; Jung, Y.M. A SERS Study of Charge Transfer Process in Au Nanorod–MBA@Cu_2O Assemblies: Effect of Length to Diameter Ratio of Au Nanorods. *Nanomaterials* **2021**, *11*, 867. [CrossRef] [PubMed]
18. Mandelbaum, Y.; Mottes, R.; Zalevsky, Z.; Zitoun, D.; Karsenty, A. Investigations of Shape, Material and Excitation Wavelength Effects on Field Enhancement in SERS Advanced Tips. *Nanomaterials* **2021**, *11*, 237. [CrossRef] [PubMed]

19. Zavyalova, E.; Ambartsumyan, O.; Zhdanov, G.; Gribanyov, D.; Gushchin, V.; Tkachuk, A.; Rudakova, E.; Nikiforova, M.; Kuznetsova, N.; Popova, L.; et al. SERS-Based Aptasensor for Rapid Quantitative Detection of SARS-CoV-2. *Nanomaterials* **2021**, *11*, 1394. [CrossRef] [PubMed]
20. Xu, M.-L.; Gao, Y.; Jin, J.; Xiong, J.-F.; Han, X.X.; Zhao, B. Role of 2-^{13}C Isotopic Glyphosate Adsorption on Silver Nanoparticles Based on Ninhydrin Reaction: A Study Based on Surface—Enhanced Raman Spectroscopy. *Nanomaterials* **2020**, *10*, 2539. [CrossRef] [PubMed]

 nanomaterials

MDPI

Article

Highly Sensitive and Stable Copper-Based SERS Chips Prepared by a Chemical Reduction Method

Pei Dai, Haochen Li, Xianzhi Huang, Nan Wang and Lihua Zhu *

School of Chemistry and Chemical Engineering, Huazhong University of Science and Technology, Wuhan 430074, China; peidai94@hust.edu.cn (P.D.); u201810449@hust.edu.cn (H.L.); huangxzh26@mail2.sysu.edu.cn (X.H.); nwang@hust.edu.cn (N.W.)
* Correspondence: lhzhu63@hust.edu.cn

Abstract: Cu chips are cheaper than Ag and Au chips for practical SERS applications. However, copper substrates generally have weak SERS enhancement effects and poor stability. In the present work, Cu-based SERS chips with high sensitivity and stability were developed by a chemical reduction method. In the preparation process, Cu NPs were densely deposited onto fabric supports. The as-prepared Cu-coated fabric was hydrophobic with fairly good SERS performance. The Cu-coated fabric was able to be used as a SERS chip to detect crystal violet, and it exhibited an enhancement factor of 2.0 × 106 and gave a limit of detection (LOD) as low as 10–8 M. The hydrophobicity of the Cu membrane on the fabric is favorable to cleaning background interference signals and promoting the stability of Cu NPs to environment oxidation. However, this Cu SERS chip was still poor in its long-term stability. The SERS intensity on the chip was decreased to 18% of the original one after it was stored in air for 60 days. A simple introduction of Ag onto the clean Cu surface was achieved by a replacement reaction to further enhance the SERS performances of the Cu chips. The Ag-modified Cu chips showed an increase of the enhancement factor to 7.6 × 106 due to the plasmonic coupling between Cu and Ag in nanoscale, and decreased the LOD of CV to 10–11 M by three orders of magnitude. Owing to the additional protection of Ag shell, the SERS intensity of the Cu-Ag chip after a two-month storing maintained 80% of the original intensity. The Cu-Ag SERS chips were also applied to detect other organics, and showing wide linearity range and low LOD values for the quantitative detection.

Keywords: surface-enhanced Raman scattering; copper; SERS chip; chemical reduction; hydrophobicity

Citation: Dai, P.; Li, H.; Huang, X.; Wang, N.; Zhu, L. Highly Sensitive and Stable Copper-Based SERS Chips Prepared by a Chemical Reduction Method. *Nanomaterials* **2021**, *11*, 2770. https://doi.org/10.3390/nano11102770

Academic Editors: Ronald Birke and Bing Zhao

Received: 17 September 2021
Accepted: 12 October 2021
Published: 19 October 2021

Publisher's Note: MDPI stays neutral with regard to jurisdictional claims in published maps and institutional affiliations.

1. Introduction

Surface-enhanced Raman spectroscopy (SERS), as a highly sensitive vibrational spectroscopy [1], is one of the most commonly used on-field spectroscopic detection techniques with advantages of convenience, rapidness and high sensitivity [2]. Typical SERS active substrates are noble metals (Au, Ag, Cu) at nanoscale. A laser can excite their localized surface plasmons resonances (LSPR) and amplify electromagnetic fields, finally leading to the SERS with electromagnetic enhancement mechanism (EM) [3,4]. Among these noble metals, Ag-based substrates produce the strongest Raman enhancement and Au-based substrates have the best long-term stability in air, whereas Cu-based substrates attract the least attention because of their insignificant SERS enhancement and poor stability, although Cu is much cheaper.

Due to the poorer LSPR of Cu, its electromagnetic enhancement is weaker than that of Ag and Au. The SERS enhancement factor obtained for simple copper substrates were reported to be 10^3–10^7 [5,6], being lower than that of silver (10^6–10^{14}) and gold (10^4–10^9) [7,8]. In order to improve the SERS performances of Cu substrates, two strategies are often employed. One is to construct the SERS substrates with compact arrangement of nanoparticles. Kowalska et al. used high pressure for the decomposition of copper hydride (CuH) to prepare SERS platforms with uniformly distributed copper nanocrystals;

the obtained substrates showed good SERS performance with enhancement factor up to 10^6–10^7 [9]. Rao et al. assembled spherical nanocopper into three-dimensional nanoporous Cu leaves by modifying Cu NPs with isooctane and polyethylene glycol, and found that the porous 3D structure gave an enhancement factor of 1.2×10^6 [10]. The other is to combine copper with other materials [11]. Dizajghorbani-Aghdam et al. co-deposited Cu NPs with a graphitic carbon nitride (gCN) support, and found that the Cu/gCN hybrids showed strong absorption in the visible light to near-IR range, resulting in an enhancement factor of 10^7 [12].

Copper is chemically active in air, and hence Cu NPs are easy to aggregate irregularly and to be oxidized, leading to the decrease of the enhancement effect. Loading Cu NPs onto the fixed substrates, such as Si wafer [13,14], glass [15], polystyrene spheres [16] and graphene oxide (GO) [17], could prevent the random aggregation of colloidal copper. A compact arrangement of Cu NPs may produce abundant hot spots with fairly high enhancement effect with improved SERS performances. Surface coating of noble metal nanoparticles is a common method to prevent oxidation [18,19]. The size and morphology of nanocopper can be controlled by modifying surfactants on the surface of copper, and the obtained coating can also hamper the aggregation and oxidation of Cu NPs. Zhang et al. encapsulated Cu NPs in a graphene shell (thickness: 1 nm) to improve the stability of Cu NPs [20]. Compared with traditional SERS coatings of metallic NPs, such as SiO_2 [21], polymers [22] and amorphous carbon [23], the few layers of graphene could strengthen the plasmonic coupling between graphene and Cu [24], exhibiting an enhancement factor of 1.15×10^6.

Copper-based SERS chips have been prepared with various methods though chemical reduction [25,26], electrochemical deposition [27], aerosol direct writing [28], subsequent dealloying process [29], magnetron sputtering [30], laser ablation [12] and chemical vapor deposition (CVD) [20]. Among these methods, the chemical and electrochemical reduction processes are ubiquitous and convenient to operate without the needs of high temperature, high level of vacuum and expensive instruments. However, the surface of Cu NPs produced by solution chemistry techniques may be contaminated by the added stabilizing agent, such as polyvinyl pyrrolidone [31], sodium dodecylbenzene sulphonate [32] or cetyltrimethylammonium bromide [33], the residues of which easily lead to disturbances in high-sensitivity detection.

In the present work, we successfully developed the Ag-modified Cu SERS chip with high sensitivity and stability by a chemical reduction method. By loading Cu NPs on the fabric, the random aggregation of colloidal copper was avoided. The compact arrangement of Cu NPs leads to the hydrophobicity and sensitivity of the copper-based SERS chip. Without any modifier to hinder SERS performance and create interference signals, the Cu substrate had a clean background. Silver could be directly deposited onto the surface-clean Cu-coated fabric by a replacement reaction. The introduction of Ag further strengthened the plasmonic coupling between Cu and Ag, therefore contributing to an improved sensitivity. The hydrophobicity of the copper membrane was of great importance because a hydrophobic surface could improve the stability of materials [34]. Being benefit from the hydrophobic effect of the hydrophobic surface and the protection from the thin silver shell, the as-prepared SERS chips had good stability in air. We should note that it looked strange but is very interesting that the SERS performances of the mildly oxidized SERS substrate could be partially recovered by a vacuum deoxygenation treatment. Owing to the contribution of chemical enhancement mechanism (CM) in this Cu-Ag chip, the target molecules with required energy levels could be selective enhanced and quantitative detected through the photo-induced charge transfer (PICT).

2. Experimental Section

2.1. Reagents

$N_2H_4 \cdot H_2O$ (85%), H_2O_2 (30%), crystal violet (CV) and ethanol were purchased from Sinopharm Chemical Reagent Co., Ltd (Shanghai, China). $Cu(CH_3COO)_2 \cdot H_2O$ and $AgNO_3$

were purchased from Shanghai Lingfeng Chemical Reagents Co., Ltd (Shanghai, China). Paraquat (PQ) was purchased from Macklin (Shanghai, China). Sibutramine hydrochloride (SH) was obtained from the National Institute for the Control of Pharmaceutical and Biological Products (Beijing, China). All the chemicals were analytically pure and used as received without further purification. Deionized water was used throughout the experiments.

2.2. Preparation of Flexible Hydrophobic Ag Modified Copper SERS Chips

The flexible SERS chips were chemically prepared by depositing hydrophobic Cu membrane on fabrics and then coating sliver on the Cu membrane through a replacement reaction. In the first step, copper was coated on fabrics with a modified chemical deposition procedure as reported [35]. Fabrics were cut into pieces with sizes of 2 cm × 2 cm, followed by fully washing with water and ethanol. After being dried, the cleaned fabric was fully immersed in a solution of copper acetate (20 mL, 10 g L^{-1}) for several minutes. Then, 800 μL of 80% hydrazine hydrate was added to it drop by drop under mild stirring. The reaction in the mix solution was kept at room temperature for 6 h, which allowed the deposition of copper particles onto the fabric. Afterward, the prepared Cu-coated fabric was taken out and washed with water and ethanol, followed by vacuum drying at 60 °C for 1 h. The obtained Cu membrane on the fabric was confirmed to be hydrophobic. In the present work, this sample was also used as a SERS chip, being referred to as a hydrophobic Cu chip.

In the second step, Ag particles were further deposited on the above obtained Cu-coated fabric by immersing it in a solution of $AgNO_3$ (0.01 M) in ethanol for 1 min. Then, the fabric was washed with water and ethanol, and vacuum dried at 60 °C for 1 h. This product was flexible hydrophobic Ag modified copper SERS chips, being referred to as a hydrophobic Cu-Ag chip. If not specified elsewhere, these two types of the chips were used in the SERS measurements after they were carefully cut into small pieces with a size of 0.5 cm × 0.5 cm.

2.3. Characterization

Scanning electron microscopy (SEM) characterization was performed on a GeminiSEM 300 (ZEISS, Heidenheim, Germany). X-ray diffraction (XRD) patterns were recorded by a SmartLab-SE diffractometer (Rigaku, Tokyo, Japan). Contact angles were obtained on an OCA20 optical contact angle meter (Dataphysics, Filderstadt, Germany). UV–vis diffuse reflectance spectra were obtained by a UV-3600 UV-VIS-NIR spectrophotometer (Shimadzu, Kyoto, Japan) in diffuse reflectance mode. XPS analysis was conducted on a K-Alpha X-ray photoelectron spectrometer (Thermo Scientific, Carlsbad, CA, USA).

2.4. SERS Detection

The specified SERS chip was immersed in the ethanol solution of analytes for 10 min, and then it was taken out for SERS detection. The SERS detection was performed with a ATR8100 portable Raman spectrometer (Optosky, Xiamen, China). A 785 nm laser was used as an excitation source with a laser power of 50 mW, and the exposure time of 5 s was set without accumulation. During the measurement, the fabric support in the chip was kept wet by adding ethanol. The detection was performed five times at different positions on the tested SERS chip, and the averaged spectrum was used for further analysis.

The Raman enhancement factor was calculated with the following equation:

$$EF = \frac{I_{SERS} \times N_{NR}}{N_{SERS} \times I_{NR}} \quad (1)$$

where I_{SERS} and I_{NR} are the intensity of the same peaks position in the SERS spectrum and normal Raman spectrum, respectively, N_{SERS} represents the number of adsorbed molecules in the SERS analysis and N_{NR} is the number of molecules in the scattering volume.

2.5. Stability Experiment

To evaluate the oxidation resistance of the substrate, a newly made SERS chip was firstly used to detect the original SERS signal of CV (I_0). Then, it was stored in the air for different periods of time and then measured the SERS intensity (I_t). The time course of the I_t/I_0 value was used to evaluate the stability of the chip for its endurance to the oxidation of substrate.

To recover of the SERS performance of the oxidized SERS substrate, the "oxidized" chip was vacuum dried at 60 °C for 2 h. The vacuum dried "recovered" chip was used for the SERS detection. Here, the so-called "oxidized" chip was obtained by immersing the "freshly prepared" chip for about 30 min in the ethanol solution of 0.01 M H_2O_2, which was prepared by dissolving a required amount of H_2O_2 (30%) into ethanol. After it was washed with ethanol and air dried, it was used to measure the SERS performance.

3. Results and Discussion

3.1. Characterizations of the Cu and Cu-Ag Chips

In the Cu chip, the fabric was used as the carrier to locate copper nanoparticles and prevent the spontaneous aggregation of them, while Cu NPs were evenly deposited on the fabric by a simple chemical reduction method with hydrazine hydrate as a reducing agent. The as-prepared Cu-coated fabric chip had a deep red color with copper evenly deposited on the filaments. The SEM image of the Cu coating on the fabric (Figure 1a–c) showed that the amount of deposited copper increased and the Cu NPs became bigger with a prolonged deposition time. For the Cu coating with deposition time of 6 h (being referred to as 6 h Cu), the Cu NPs with sizes of about 100–200 nm were packed together closely (Figure 1b). The packing-induced micro-nano structure of the deposited Cu NPs made the surface of the fabric chip hydrophobic, yielding a contact angle of 144.0° (Figure 1d) for 6 h Cu. The EDS mapping of Cu on the Cu coating (Figure 1i) confirmed the even distribution of Cu on the surface of the fabric.

After the 6 h Cu sample was treated by immersion in a solution of AgNO3 (0.01 M) in ethanol, the Cu chip was converted to the Cu-Ag chip, which was similarly characterized. The SEM image (Figure 1e–g) demonstrated that the Ag NPs covered on Cu NPs generated from the discrete small particles to the monolithic structure. As shown in Figure 1e, the aggregated Cu NPs were covered by much smaller particles on the Cu-Ag coating with 1 min replacement time (6 h Cu-1 min Ag). This is possibly related to the following processes: the replacement reaction led to the generation of Ag NPs with smaller particle sizes and also thinned the out-most Cu particles through the partial dissolution of Cu. Due to the deposition of Ag, the fabric chip became dark in color, indicating the deposition of Ag NPs with small sizes. The EDS mapping of Cu and Ag on the Cu-Ag chip (Figure 1j,k) confirmed the even distribution of both Cu and Ag on the surface of the fabric. The deposition amount of Ag was little, and atomic ratio of Cu: Ag was around 7:1. The easy detection of Cu under the Ag layer by EDS mapping also signed that the outermost layer of Ag was very thin. This thin layer did not change the hydrophobic nature of the pre-obtained Cu layer, and hence the contact angle on the surface of the Cu-Ag chip was still as large as 141.5° (Figure 1h). The relationship between SERS performance and structural morphology of Cu and Cu-Ag chip will be discussed later.

The crystal structures and chemical structures of the Cu and Cu-Ag chips were investigated by XRD and XPS analysis. As shown in Figure 2a, the XRD patterns of the Cu coating showed two peaks at 43.3° and 50.5°, which matched well with the diffraction peaks of (1 1 1) and (2 0 0) for the Cu metal (JCPDS 04-0836). Figure 2b showed Cu 2p XPS spectra of the copper coating. There were only two spin-orbit splitting components of 2p1/2 (952 eV) and 2p3/2 (932 eV) without the satellite peaks around 943 eV, further indicating the state of metallic copper.

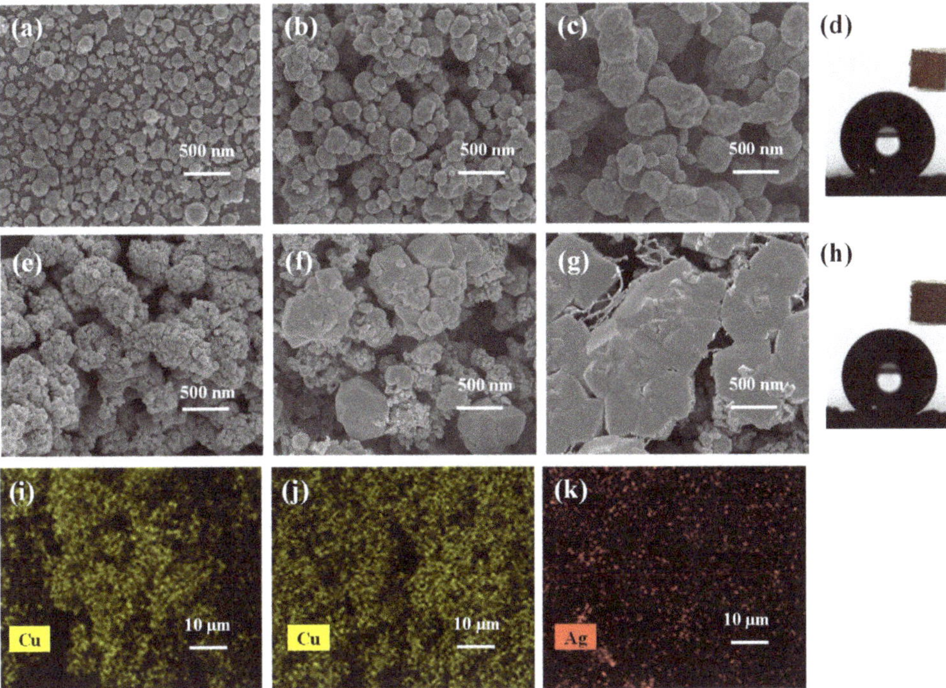

Figure 1. Characterization of Cu and Cu-Ag chips. (**a–c**,**e–g**) SEM images of the Cu coating with Table 1. h, (**b**) 6 h, (**c**) 12 h and Cu-Ag coating with the replacement time of (**e**) 1 min, (**f**) 3 min and (**g**) 5 min. (**d**,**h**) Contact angles of the coatings of 6 h Cu (**d**) and 6 h Cu-1 min Ag (**h**) on the fabric together with their photos in the inset. (**i**) EDS mapping of Cu on the Cu coating. (**j**,**k**) EDS mapping of Cu (**j**) and Ag (**k**) on the Cu-Ag coating.

Table 1. Comparison between copper-based SERS chips prepared by various methods.

SERS Substrate	Preparation Method	Targets	LOD (M)	EF	Ref.
Cu/gCN	pulsed laser ablation	CV R6G	10^{-7} 10^{-6}	7.2×10^7 1.3×10^7	[12]
Cu NPs/Si wafer	Si–H bond assembly	R6G	10^{-9}	2.3×10^7	[14]
Cu-doped glass	thermal annealing	RhB	10^{-9}	1.5×10^8	[15]
Cu NP arrays	ion-sputtering deposition	4-ATP	10^{-7}	1.6×10^7	[16]
Cu@G-NGNs	chemical vapor deposition	R6G	10^{-7}	1.1×10^6	[20]
Cu NPs	aerosol direct writing	RhB	10^{-6}	2.1×10^5	[28]
nanoporous Cu	subsequent dealloying	R6G	10^{-9}	4.7×10^7	[29]
Cu nanoislands	magnetron sputtering	4-ATP	10^{-7}	4.0×10^4	[30]
mesoporous Cu films	electrochemical deposition	R6G	10^{-6}	3.8×10^5	[27]
Cu/rGO	chemical reduction with rGO as stabilizing agent	CV	/	/	[17]
Cu NPs	chemical reduction with gelatin as stabilizing agent	CV	/	3.6×10^3	[25]
Cu NPs	chemical reduction with octadecylamine as stabilizing agent	RhB	/	8.6×10^3	[26]
3D nanoporous Cu leaves	chemical reduction with isooctane/PEG as stabilizing agent	4-MBA	/	1.2×10^6	[11]
Cu-coated fabric Cu-Ag-coated fabric	chemical reduction without any stabilizing agent	CV	10^{-8} 10^{-11}	2.0×10^6 7.6×10^6	This work

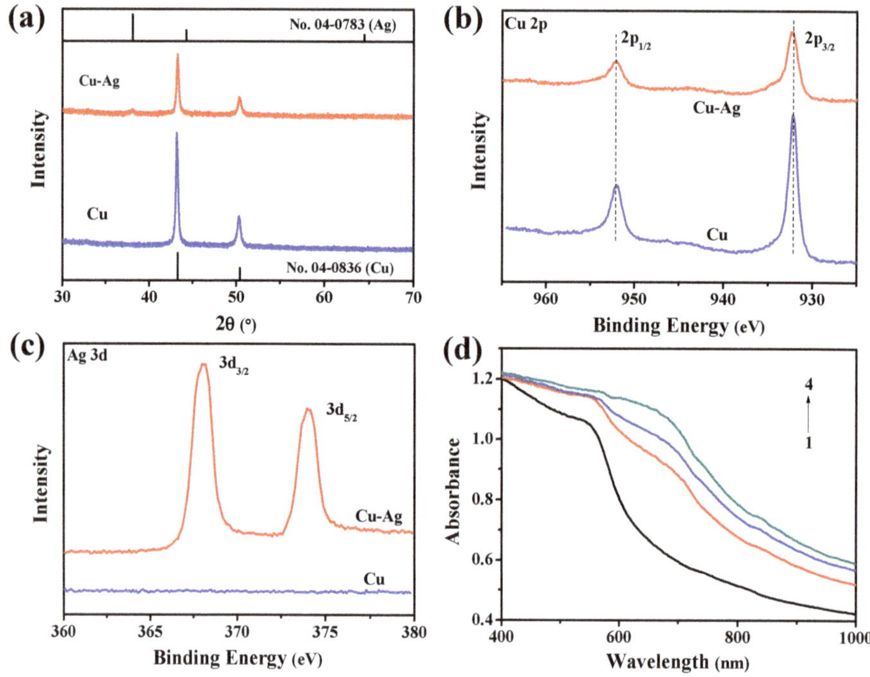

Figure 2. XRD patterns (**a**), XPS high resolution spectra of Cu 2p (**b**), Ag 3d (**c**) of the Cu and Cu-Ag chips. (**d**) VIS-NIR diffuse reflectance spectra of Cu-based fabrics with the Ag replacement Table 1. 0, (2) 1, (3) 3 and (4) 5 min in the preparation.

The XRD pattern showed that a new diffraction peak appeared at 38.3° (Figure 2a), corresponding to the (1 1 1) crystal plane of silver (JCPDS 04-0783). As shown in the Cu 2p XPS spectra (Figure 2b), the two characteristic peaks of metallic copper at 932 eV (Cu $2p_{3/2}$) and 952 eV (Cu $2p_{1/2}$) were observed for the Cu-Ag chip like that in the case of the Cu chip. Due to the covering by the thin Ag layer, the intensities of these Cu peaks were decreased slightly. The XPS Ag 3d spectra (Figure 2c) of the Cu-Ag-coated fabric showed two peaks at 374 eV (Ag $3d_{3/2}$) and 368 eV (Ag $3d_{5/2}$), indicating the deposition of metallic silver.

Figure 2d showed the VIS-NIR diffuse reflectance spectra of the Cu-coated fabrics with different deposition times of Ag. The surface plasmon resonance (SPR) peak of Cu coating appeared at 556 nm, and the introduction of Ag enhanced this absorption and generated a new SPR peak in the near infrared region. Along with the growing of the Ag shell, the SPR peak in the near infrared region was blue shifted from 705 nm to 672 nm until it attenuated the SPR peak of Cu to merge into a single peak. This might influence the SERS performances of the chip, as discussed later.

3.2. SERS Performances of the Cu Chip

In the Cu chip, metallic copper nanoparticles were deposited on the fabric and generated to a hydrophobic surface, which is possibly favorable to its SERS performance. As shown in Figure 3a, the Cu deposition time (and hence the deposition amount) influenced the SERS responses of CV on the Cu chip. By using the intensity of the strongest peak of CV at 1617 cm^{-1}, the SERS signal intensity was plotted against the deposition time in Figure 3b. When increasing the deposition time from 0 to 6 h, the SERS signals of CV were rapidly increased from 0 to 1300 a.u.; further increasing the deposition time from 6 to 24 h decreased the SERS signal intensity, but finally keeping at 250 a.u. at about 24 h. The Cu deposition time mainly affected the individual Cu particles (the particle sizes, packing patterns) and the hydrophobicity of the surface. The hydrophobicity of the surface was monitored by measuring the contact angle. As shown in Figure 3b, once the Cu filmed was formed on the fabric, the contact angle was greater than 120°, showing hydrophobicity. For example, the contact angle of the chip surface with deposition time of 6 and 12 h was measured to be 144.0° and 150°, respectively. Because the deposition time dependence of the SERS signal intensity was greatly different from the deposition time dependence of the contact angle in the curve shape as shown in Figure 3b, the weak variation of the surface hydrophobicity would not be an important factor influencing the deposition time dependence of the SERS signal intensity. Therefore, the surface morphology of the chip was checked for different periods of deposition time. As shown in Figure 1a–c, the size of Cu NPs gradually increased and packed together with the prolonging of deposition time. When the deposition time is 1h (Figure 1a), the Cu NPs deposited on the fabric was small and loose with the size of 80 nm. When the deposition time was prolonged to 6 h (Figure 1b), the amount and diameter of Cu deposition on the fabric increased significantly, a large number of Cu NPs (100–200 nm) tightly packing to form hot spots. Further extended the deposition time to 12h (Figure 1c), the small size Cu NPs merged to form large Cu NPs with the size of around 300 nm, so the surface roughness and number of hot spots decreased. Based on the above discussions, the Cu deposition time was selected at 6 h hereafter.

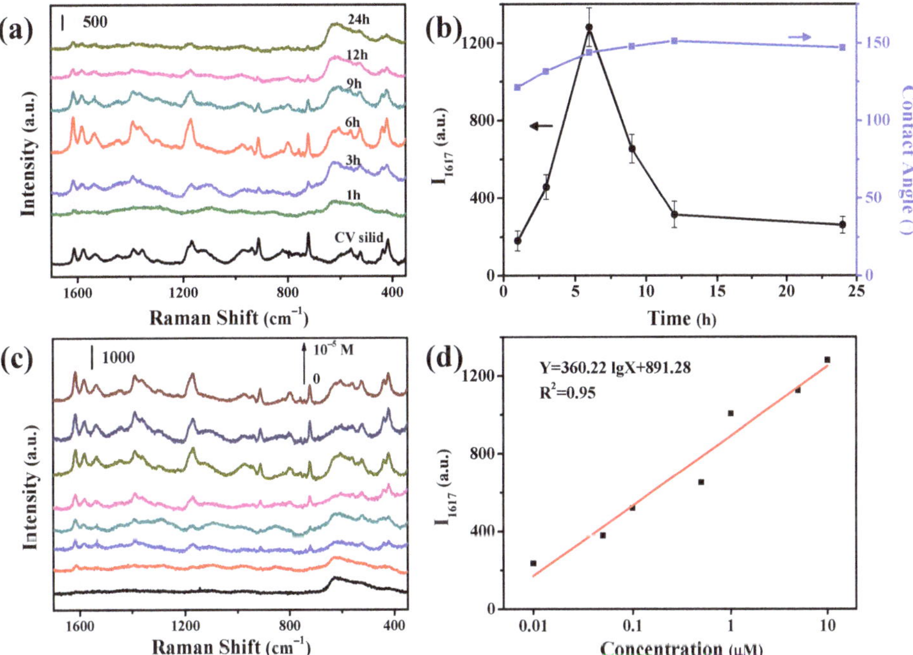

Figure 3. SERS performances of the Cu chip. (**a**) SERS spectra of 1 ppm CV on the Cu chips prepared with the different deposition time. (**b**) Influences of the Cu deposition time on the peak intensity of CV at 1617 cm^{-1} and the contact angle of the chip surface. (**c**) SERS spectra of CV at various concentrations on the Cu chips. (**d**) A plot of the peak intensity of CV at 1617 cm^{-1} against CV concentration.

By using the sharp peak of CV at 1617 cm^{-1}, the enhancement factor (EF) of the Cu chip was evaluated to be 2.0 × 106. Table 1 compared the EF values of the copper-based SERS chips prepared by various methods. This comparison indicated that the SERS performance of the presently developed Cu chip was somewhat poorer than that prepared by expensive physical methods (about 107), but much higher than that prepared by other chemical reduction methods (103–105). This was possibly related to the slow deposition of

Cu in the present work, leading to more tightly packing of Cu NPs for the generation of more hot spots in the chip. In our method, no any organic stabilizing agents were used, and hence the Cu-coated fabric exhibited a clean background without interference peaks (Figure 3c). We measured the SERS spectra of CV at different concentrations (Figure 3c), and plotted the peak intensity at 1617 cm^{-1} against the CV concentration (Figure 3d). It was found that the SERS intensity was linearly correlated with the logarithm of CV concentration in the range of 10^{-8}–10^{-5} M, with a LOD of 10^{-8} M.

3.3. *SERS Performances of the Cu-Ag Chip*

As shown in Table 1, the SERS performance of the presently prepared Cu chip needs to be promoted further in comparison with the best ones. More importantly, we found that the SERS performance of the Cu chip was decreased greatly after it was stored in air for several weeks due to its poor oxidation resistance (more details will be described in the next section). Therefore, it was required to further increase the SERS enhancing effect and the oxidation resistance of the SERS substrate. As we know, metallic Ag has stronger intrinsic SERS effect and is much more inert in air than metallic Cu. Therefore, our strategy was to cover a very thin layer of nano-Ag on the Cu chip by immersion plating.

As shown in Figure 4a, as the replacement time of Ag was prolonged from 0 to 1 min, the Ag-modified Cu chip yielded a fast increasing of the SERS signal intensity (at 1617 cm^{-1}) from 1281 to 4275 a.u. together with an increase of the enhancement factor to 7.6 × 106. This was explained by considering the effects of the deposited Ag: the deposited Ag NPs increased the surface roughness (as confirmed by the SEM observation in Figure 1e), being favorable to producing more hot spots; the initially deposited Ag NPs had a SPR peak at 705 nm, which matched better with the laser excitation of 785 nm. When the replacement time of Ag was more than 3 min, the SPR peak blue shifted, and its SERS performance decreased. This may because that the Ag covered the Cu, shielding the contribution of inner Cu. Figure 1e–g showed the SEM image of Cu-Ag chip with the replacement time of 1–5 min. With the prolonging of replacement time of Ag from 1 to 5 min, the Cu further dissolved, the Ag became larger to completely coat the Cu NPs (Figure 1f) and finally generated the Ag particles in micro scale (Figure 1g). Beyond that, Ag with higher surface energy was more easily wetted by water than Cu. Therefore, the contact angle of the Cu-based chip decreased to 141.5° with the increase of silver content, and finally tended to be stable due to the Ag completely covering the Cu (Figure 4a). Therefore, in the present work, the Ag deposition time was selected at 1 min for preparing the Cu-Ag chip.

The SERS spectra of CV at various concentrations were recorded on the Cu-Ag chip as shown in Figure 4b. In comparison with the spectra recorded on the Cu chip for the specified individual concentrations of CV (Figure 3c), the spectra recorded on the Cu-Ag chip were much enhanced in the peak intensity. From the plot of the peak intensity at 1617 cm^{-1} against the CV concentration (Figure 4c), it was found that there was a linear correspondence between the SERS intensity and the logarithm of CV concentration in the range of 10^{-9}–10^{-6} M, with an LOD as low as 10^{-11} M.

In order to evaluate the uniformity of the as-prepared SERS chips, 50 points on the same Cu-Ag chip were selected randomly to detect the SERS signals. As shown in Figure 4c, the signal intensity of CV, especially for its strongest peak at 1617 cm^{-1}, was very close to each other. Furthermore, we acquired the SERS spectra of CV at 1 $mg{\cdot}L^{-1}$ on ten Cu-Ag chips as shown in Figure 4d, and found that these spectra were also very close to each other with the relative standard deviation of only 12.6%. These demonstrated that the presently developed Cu-Ag chips have good reproducibility in term of both intra- and inter-batches.

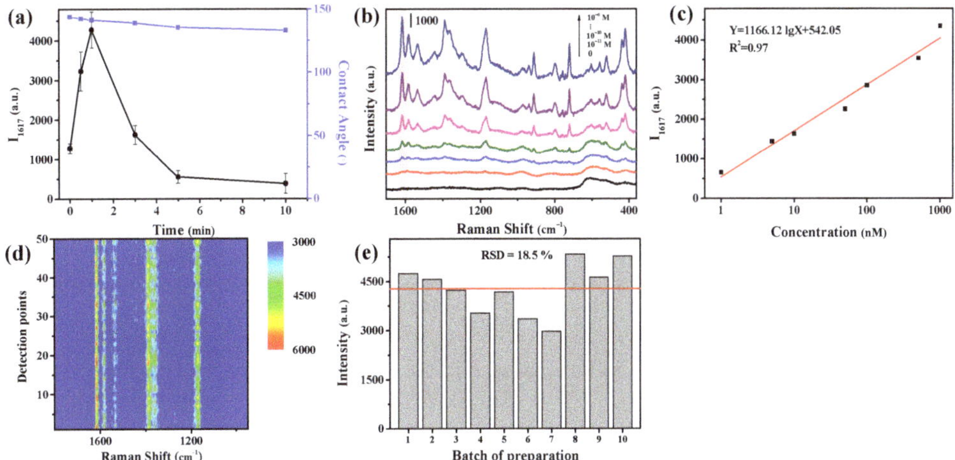

Figure 4. SERS performance of the Cu-Ag chip. (**a**) Effects of the replacement time of sliver on the peak intensity of CV at 1617 cm^{-1} and the contact angle of the chip surface. (**b**) SERS spectra of CV at various concentrations on the Cu-Ag chip. (**c**) A plot of peak intensity of CV at 1617 cm^{-1} on the Cu-Ag chip against CV concentration. Reproducibility of SERS signals obtained on the (**d**) intra-batch and (**e**) inter-batch Cu-Ag chips.

3.4. Stability of the Cu-Based SERS Chips

The resistance of copper to air oxidation is not so good, and the oxidative corrosion of Cu may decrease the SERS performances of the Cu chips. By using the characteristic peak of CV at 1617 cm^{-1} as a reference, we recorded the original peak intensity on the newly prepared chip (I0) and the peak intensity (It) after the chip was stored in the air for a specified period time of t, and then used the ratio of It/I0 to evaluate the resistance of the chip to air oxidation. Here, three different chips were tested: the first chip was a Cu chip obtained with a Cu deposition time of 3 h, being referred to as Cu(3 h) chip, the surface of which had a contact angle of 131.7°; the second was a Cu chip obtained with a Cu deposition time of 6 h, being referred to as Cu(6 h) chip, the surface of which had a contact angle of 144.0°; the third was a Cu-Ag chip that was obtained by depositing Ag for 1 min on the Cu(6 h) chip, and it showed a contact angle of 141.0°. As shown in Figure 5, on the Cu(3 h) chip, the deposited Cu particles on the fabric were less densely packed and less hydrophobic, and the air oxidation during the storage of the chip was serious, which led to a fast decrease of the It/I0 ratio, only 6% of the response ability was kept after storing 60 days. In contrast, due to the much improved deposition of Cu, the decrease of the I_t/I_0 ratio on the Cu (6 h) chip became considerably slower. For the Cu-Ag SERS chip, its SERS performance maintained about 80% after storing in air for 60 days. It indicated that the Cu-Ag chip with good hydrophobicity exhibited high oxidation resistance.

We also evaluated the stability of the SERS substrate by using an ethanol solution of H_2O_2 (0.01 M) to accelerate the oxidation of the substrate as shown in Figure 6a (curve 1). With the increase of oxidation time, the SERS signal intensity (in term of the I_t/I_0 ratio) of the as-prepared SERS chip (i.e., the oxidized chip) was rapidly decreased. When the oxidation time was 30 min, the relative intensity of CV was decreased to 4% of the original one. It was very interesting that the great loss of the oxidized chip in the SERS signaling could be substantially recovered by an after-treatment through vacuum drying as shown in Figure 6a (curve 2). For example, the vacuum drying treatment could recover the relative SERS intensity of the chip being oxidized for 30 min from 4% to about 79%.

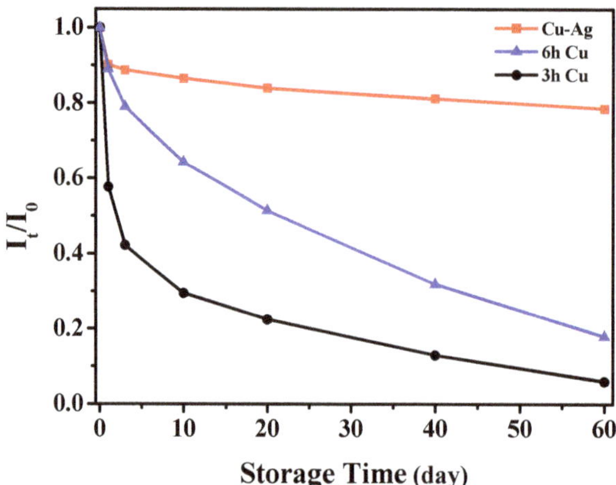

Figure 5. Endurances of SERS performances of various SERS chips to the air oxidation during storage: Cu(3 h) chip, Cu(6 h) chip and Cu-Ag chip.

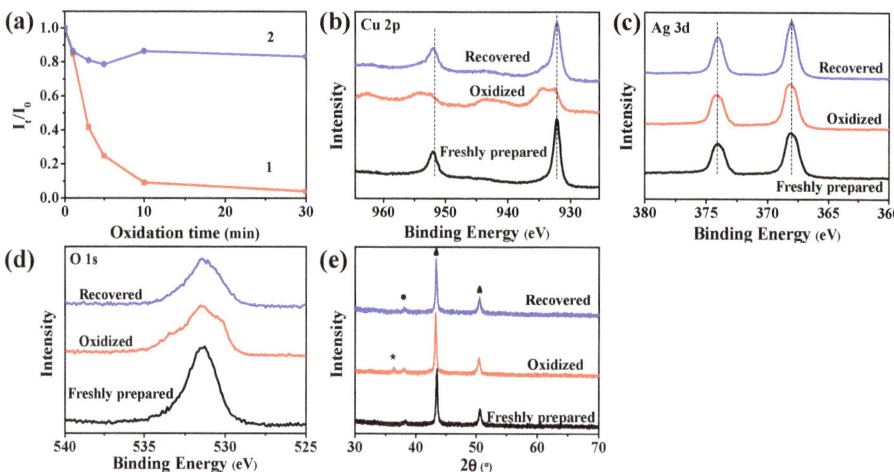

Figure 6. Recovering of the oxidized Cu-Ag SERS chip. (**a**) Dependence of the relative SERS intensity (I_t/I_0) of the Cu-Ag chip on oxidation time (curve 1) and the recovered value after a vacuum drying treatment of the oxidized chip (curve 2). (**b–d**) XPS spectra of the freshly prepared, oxidized and recovered Cu-Ag chips: (**b**) Cu 2p, (**c**) Ag 3d, and (**d**) O 1s envelops. (**e**) XRD patterns of the freshly prepared, oxidized and recovered Cu-Ag chips.

To understand what happened during the oxidation and the recovering treatment, we characterized the freshly prepared, oxidized and recovered Cu-Ag chips by using XPS and XRD (Figure 6b–e). As with that being discussed for Figure 2b, in the Cu 2p XPS spectrum of the freshly prepared Cu-Ag chip there were two characteristic peaks of metallic copper at 952 eV (Cu $2p_{1/2}$) and 932 eV (Cu $2p_{3/2}$), both of which being attributed to metallic Cu. However, in the spectrum of the oxidized chip (Figure 6b), the two peaks at 952 eV and 932 eV broadened towards the high binding energy region, indicating part of he copper converted to Cu(II), and two satellite peaks of Cu(II) appeared at 943 eV and 963 eV. These suggested that the oxidation induced the generation of Cu(II) as the oxidation product. The recovering treatment made the spectrum of the oxidized chip almost the same as that of the freshly prepared Cu-Ag chip. This meant that the recovering treatment by vacuum drying eliminated the oxidation product Cu(II) species.

Similarly, as discussed for the XPS Ag 3d spectrum of the fresh Cu-Ag chip in Figure 2c, the two peaks at 374 eV (Ag $3d_{3/2}$) and 368 eV (Ag $3d_{5/2}$) were attributed to the deposited metallic silver. The oxidation of the chip did not induce observable changes in the Ag 3d XPS spectrum, and the recovering treatment yielded yet no observable changes in the Ag 3d XPS spectrum (Figure 6c). These hinted that neither the oxidation-induced loss of the chip in the SERS sensing nor its recovering was directly related to the deposited Ag particles.

As shown in Figure 6d, in the XPS O 1s spectrum of the fresh Cu-Ag chip, a peak was observed at 532 eV, which was attributed to organic C-O bonding of fabric. After the chip was oxidized for 30 min, this peak became much broader, and its decovolution indicated that the oxygen was attached on the metal of Cu-Ag chip after oxidization, among these peaks, those at 534 eV and 530 eV were assigned to the adsorbed oxygen and metal oxides, respectively [36,37]. After the recovering treatment, both the peaks at 534 eV and 530 eV were much depressed in the intensity. This suggested that the oxygen attached on metal was partly removed.

The XRD patterns of the freshly prepared Cu-Ag chip are displayed in Figure 2a and were discussed before. After the oxidation, the oxidized Cu-Ag chip exhibited a new peak from cupric oxide as shown in Figure 6e. The new peak at 36.4° was assigned to the diffraction peak of (1 1 1) for the $Cu_{2+1}O$ (JCPDS 05-0667). $Cu_{2+1}O$ was Cu_2O with metal excess defects, indicating that the O atom of it was easy to lose and lead to the oxygen vacancy [38,39]. A substantial part of the oxygen could be removed by the vacuum drying treatment. As confirmed by the XRD analysis, the diffraction peak of $Cu_{2+1}O$ for the oxidized chip was disappeared in the XRD patterns of the recovered Cu-Ag chip. These results showed that the O atom of surface with low binding energy could be removed by vacuum drying, which substantially recovered the SERS performance of the oxidized Cu-Ag SERS chip.

3.5. *Quantitative SERS Detection of Other Organic Compounds with the Cu-Ag Chip*

As discussed in Section 3.3 "SERS performances of the Cu-Ag chip", a SERS method was developed by using the Cu-Ag chip for the detection of CV, which gave a linear correspondence between the SERS intensity and the logarithm of CV concentration in the range of 10^{-9}–10^{-6} M, with a LOD as low as 10^{-11} M (Figure 4b,c). Here, we used this chip to detect more organic compounds to demonstrate its generality as a good SERS sensor.

As shown in Figure 7a, the SERS spectra of nine organic compounds with relatively strong Raman activity were acquired at 1 ppm. It was found that five of them could be detected on the Cu-Ag chip, including sibutramine hydrochloride (SH), paraquat (PQ), rhodamine b (RhB), crystal violet (CV) and methylene blue (MB), which were classified as Group 1. In contrast, the other four (cysteine (Cys), saccharin sodium (SS), melamine (MEL) and 4-mercaptobenzoic acid (4-MBA)), being classified as Group 2, could not be detected on this chip. This difference indicated that the Cu-Ag chip had selective chemical enhancement for different targets due to CM. The CM of Raman scattering was ascribed to the chemical interaction between metal-molecule, whereby the charge transfer (CT) in the metal-molecule system was supposed to alter the electron density distribution of molecules, resulting in greater polarizability and thus enhanced Raman scattering. The energy levels of HOMO/LUMO of the above targets were calculated by Gaussian method, and it was found that the minimum energy barrier for the charge transfer between metal and molecules in Group 1 (Figure 7b) were in the range of 0.54–1.31 eV, being considerably smaller than those of the molecules in Group 2 in the range of 2.03–4.66 eV (Figure 7c). Therefore, a charge transfer is easily induced by irradiating 785 nm laser, leading to the photo-induced charge transfer (PICT). The charge transfer between the metal-molecule pair may occur in the direction of metal-to-molecule or molecule-to-metal, which depends on the Fermi level of metal and the HOMO/LUMO levels of the molecule (Figure 7d) [37–39].

In the Cu-Ag chip, the work function of Cu (4.65 eV) is higher than that of Ag (4.3 eV), and the electrons will transfer from Ag to Cu until the energy thermodynamic equilibrium level. Moreover, the Fermi level of Ag will be descended, and that of Cu will be raised up to attain the equilibrium level of −4.475 eV. To achieve PICT by the laser of 785 nm (ΔE = 1.58 eV), the target molecules should meet the requirement: HOMO level in the range of −4.475 eV to −6.055 eV, or LUMO level in the range of −2.895 eV to −4.475 eV. The molecules in Group 2 could not meet the above requirement, and hence there were no characteristic peaks in their SERS spectra obtained on the Cu-Ag chip even at a concentration of 100 ppm, while the molecules in Group 1 meet the requirement and could all be detected at 1 ppm. Moreover, the stronger charge transfer will be achieved as the energy level is closer to Fermi level of Cu-Ag chip. Due to the similar EM of substrate for adsorbed targets, the larger EF typically was achieved when the energy level of molecule was closer Fermi level. For example, the EF value of SH, MB, CV and RhB was 3.7×10^7, 2.9×10^7, 7.6×10^6 and 1.2×10^6 generally following an decreasing order in their minimum $|\Delta E|$ between the HOMO/LUMO of the molecules and Fermi of metal (0.54 eV, 0.87 eV, 1.10 eV and 1.31 eV). This clearly explained why the Cu-Ag chip had a selectively to the detection of different target molecules.

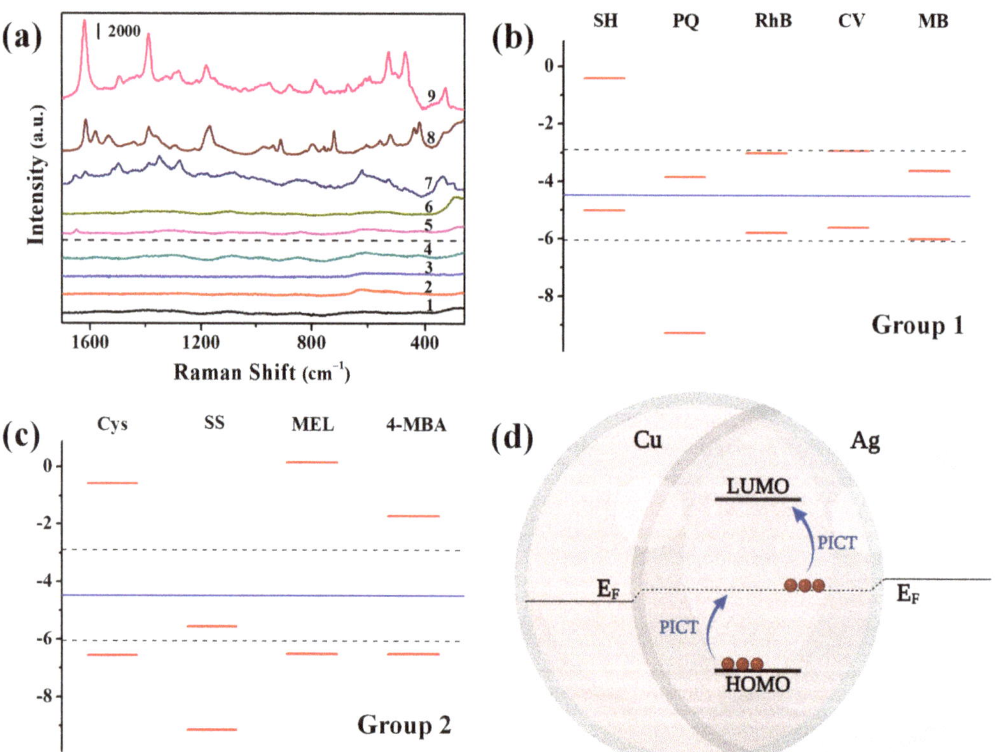

Figure 7. (**a**) SERE spectra of (1) Cys, (2) SS, (3) MEL, (4) 4-MBA, (5) SH, (6) PQ, (7) RhB (8) CV and (9) MB at 1 ppm on the Cu-Ag SERS chip. The energy level diagram of the molecules in (**b**) Group 1 and (**c**) Group 2. (**d**) The electron transition model between Cu-Ag chip and target molecules.

In comparison with CV, the SERS activity of SH and PQ were much weaker, and hence we checked their quantitative detection to further evidence the merits of the Cu-Ag chip (Figure 8). As their concentration was increased, their characteristic peaks became stronger.

By using the strongest peak of each of the tested compounds, the SERS intensity was plotted against the logarithm of its concentration. It was found that the SERS intensity was well linearly correlated with the concentration in log scale for each compound. The linear range and LOD values were evaluated to be 10^{-8}–10^{-5} M and 10^{-9} M for PQ, 10^{-6}–10^{-3} M and 10^{-7} M for SH, respectively. The wide linear ranges and low LOD values for all the tested targets indicated that the use of the Cu-Ag SERS chip provided a good quantitative detection method for SERS analysis.

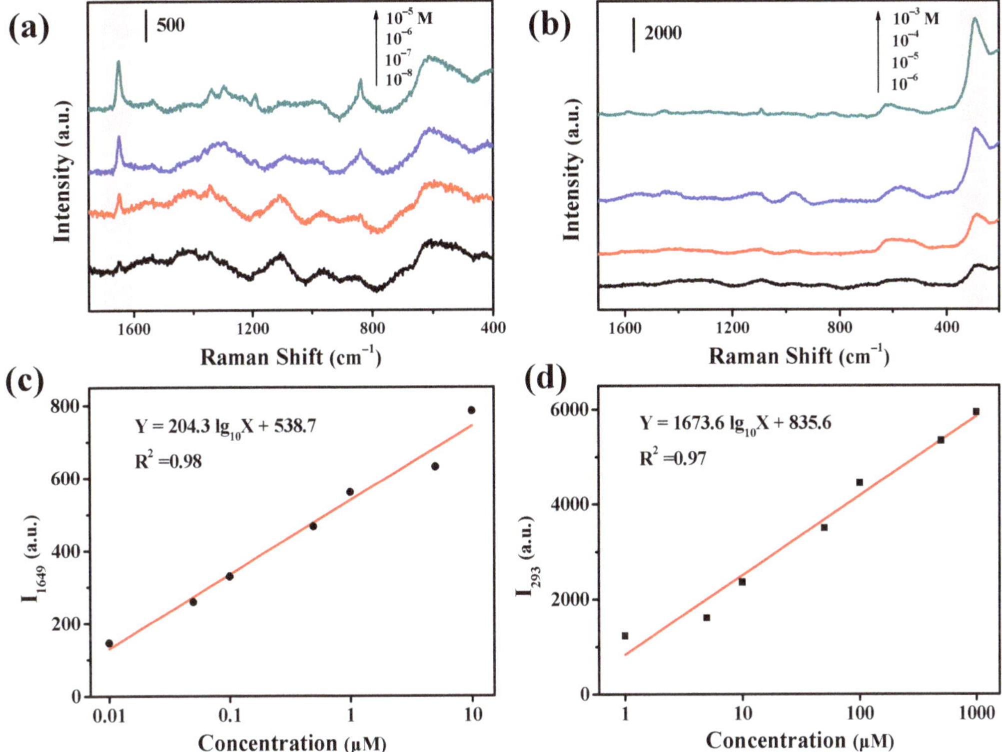

Figure 8. SERS spectra of (**a**) PQ and (**b**) SH at various concentrations on the Cu-Ag SERS chip. Linearity between the SERS intensity and $\log_{10}$ (concentration) of (**c**) PQ and (**d**) SH by using their strongest peaks.

4. Conclusions

In the present work, we have developed a sensitive and stable copper-based SERS chip by a chemical reduction method. The chemical deposition of Cu NPs on the fabric support ensured the Cu chip had no background interference of stabilizers because no stabilizers were used in the preparation. This permitted the quantitative detection of CV in a range from 10^{-8} M to 10^{-5} M with an LOD as low as 10^{-8} M. By considering the further improvement of the SERS enhancing effect and the air oxidation resistance of the Cu chip, a very thin layer of Ag NPs was deposited on the Cu chip by a replacement reaction. The introduction of Ag further enhanced the SERS performance of Cu chip, and the EF was increased from 2.0×10^6 to 7.6×10^6 after modification with Ag. Both the hydrophobicity and Ag shell improved the oxidation resistance of the Cu-Ag SERS chip, the SERS intensity kept 80% of the original signal after storing in the air for 60 days. The O atoms on the oxidized substrate could even be removed by vacuum drying to restore its SERS performance. The as-prepared Cu-Ag chip was used for detection of various targets, and results indicated that the molecules with required energy levels could be selectively detected at 1 ppm due to the CM caused by photo-induced charge transfer. The SERS method on this Cu-Ag chip also showed a wide linearity range and low LOD values for the quantitative detection of these organic compounds.

Author Contributions: Methodology and writing—original draft preparation, P.D.; investigation, H.L.; validation, X.H.; software, N.W.; project administration, L.Z. All authors have read and agreed to the published version of the manuscript.

Funding: This research was funded by the National Natural Science Foundation of China grant number 22076052, 21976063. And The APC was funded by the National Natural Science Foundation of China grant number 22076052.

Acknowledgments: The authors acknowledge the financial support from the National Natural Science Foundation of China (Grant No. 22076052, 21976063). The electron transition model diagram was created with BioRender.com.

Conflicts of Interest: The authors declare no conflict of interest.

References

1. Haynes, C.L.; McFarland, A.D.; Duyne, R.P.V. Surface-enhanced Raman spectroscopy. *Anal. Chem.* **2005**, *77*, 338 A–346 A. [CrossRef]
2. Jiang, Y.; Sun, D.; Pu, H.; Wei, Q. Surface-enhanced Raman Spectroscopy (SERS): A novel reliable technique for rapid detection of common harmful chemical residues. *Trends Food Sci. Tech.* **2018**, *75*, 10–22. [CrossRef]
3. Wei, H.; Xu, H. Hot spots in different metal nanostructures for plasmon-enhanced Raman spectroscopy. *Nanoscale* **2013**, *5*, 10794–10805. [CrossRef] [PubMed]
4. Tan, T.; Tian, C.; Ren, Z.; Yang, J.; Chen, Y.; Sun, L.; Li, Z.; Wu, A.; Yin, J.; Fu, H. LSPR-dependent SERS performance of silver nanoplates with highly stable and broad tunable LSPRs prepared through an improved seed-mediated strategy. *Phys. Chem. Chem. Phys.* **2013**, *15*, 21034–21042. [CrossRef] [PubMed]
5. Athira, K.; Ranjana, M.; Bharathi, M.S.S.; Reddy, B.N.; Babu, T.G.S.; Rao, S.V.; Kumar, D.V.R. Aggregation induced, formaldehyde tailored nanowire like networks of Cu and their SERS activity. *Chem. Phys. Lett.* **2020**, *748*, 137390. [CrossRef]
6. Qiu, H.; Xu, S.; Chen, P.; Gao, S.; Li, Z.; Zhang, C.; Jiang, S.; Liu, M.; Li, H.; Feng, D. A novel surface-enhanced Raman spectroscopy substrate based on hybrid structure of monolayer graphene and Cu nanoparticles for adenosine detection. *Appl. Surf. Sci.* **2015**, *332*, 614–619. [CrossRef]
7. Chen, H.; Lin, M.; Wang, C.; Chang, Y.; Gwo, S. Large-scale hot spot engineering for quantitative SERS at the single-molecule scale. *J. Am. Chem. Soc.* **2015**, *137*, 13698–13705. [CrossRef]
8. Kleinman, S.L.; Sharma, B.; Blaber, M.G.; Henry, A.; Valley, N.; Freeman, R.G.; Natan, M.J.; Schatz, G.C.; Duyne, R.P.V. Structure enhancement factor relationships in single gold nanoantennas by surface-enhanced Raman excitation spectroscopy. *J. Am. Chem. Soc.* **2013**, *135*, 301–308. [CrossRef] [PubMed]
9. Kowalska, A.A.; Kaminska, A.; Adamkiewicz, W.; Witkowska, E.; Tkacz, M. Novel highly sensitive Cu-based SERS platforms for biosensing applications. *J. Raman Spectrosc.* **2015**, *46*, 428–433. [CrossRef]
10. Rao, G.; Jian, X.; Lv, W.; Zhu, G.; Xiong, J.; He, W. A highly-efficient route to three-dimensional nanoporous copper leaves with high surface enhanced Raman scattering properties. *Chem. Eng. J.* **2017**, *321*, 394–400. [CrossRef]
11. Li, Z.; Jiang, S.; Xu, S.; Zhang, C.; Qiu, H.; Li, C.; Sheng, Y.; Huo, Y.; Yang, C.; Man, B. Few-layer MoS2-encapsulated Cu nanoparticle hybrids fabricated by two-step annealing process for surface enhanced Raman scattering. *Sens. Actuators B-Chem.* **2016**, *230*, 645–652. [CrossRef]
12. Dizajghorbani-Aghdam, H.; Miller, T.S.; Malekfar, R.; McMillan, P.F. SERS-active Cu nanoparticles on carbon nitride support fabricated using pulsed laser ablation. *Nanomaterials* **2019**, *9*, 1223. [CrossRef] [PubMed]
13. Jiang, W.; Shan, W.; Ling, H.; Wang, Y.; Cao, Y.; Li, X. Surface-enhanced Raman scattering of patterned copper nanostructure electrolessly plated on arrayed nanoporous silicon pillars. *J. Phys. Condens. Matter* **2010**, *22*, 415105. [CrossRef] [PubMed]
14. Shao, Q.; Que, R.; Shao, M.; Cheng, L.; Lee, S. Copper Nanoparticles Grafted on a Silicon Wafer and Their Excellent Surface-Enhanced Raman Scattering. *Adv. Funct. Mater.* **2012**, *22*, 2067–2070. [CrossRef]
15. Pereira, A.J.; Gomes, J.P.; Lenz, G.F.; Schneider, R.; Chaker, J.A.; de Souza, P.E.N.; Felix, J.F. Facile shape-controlled fabrication of copper nanostructures on borophosphate glasses: Synthesis, characterization, and their highly sensitive Surface-Enhanced Raman Scattering (SERS) properties. *J. Phys. Chem. C* **2016**, *120*, 12265–12272. [CrossRef]
16. Ding, Q.; Hang, L.; Ma, L. Controlled synthesis of Cu nanoparticle arrays with surface enhanced Raman scattering effect performance. *RSC Adv.* **2018**, *8*, 1753–1757. [CrossRef]
17. Gill, A.A.S.; Singh, S.; Nate, Z.; Chauhan, R.; Thapliyal, N.B.; Karpoormath, R.; Maru, S.M.; Reddy, T.M. A novel copper-based 3D porous nanocomposite for electrochemical detection and inactivation of pathogenic bacteria. *Sens. Actuators B-Chem.* **2020**, *321*, 128449. [CrossRef]

18. Ouyang, L.; Wang, Y.; Zhu, L.; Irudayaraj, J.; Tang, H. Filtration-assisted fabrication of large-area uniform and long-term stable graphene isolated nano-ag array membrane as surface enhanced Raman scattering substrate. *Adv. Mater. Interfaces* **2018**, *5*, 1701221. [CrossRef]
19. Ban, R.; Yu, Y.; Zhang, M.; Yin, J.; Xu, B.; Wu, D.; Wu, M.; Zhang, Z.; Tai, H.; Li, J.; et al. Synergetic SERS enhancement in a metal-like/metal double-shell structure for sensitive and stable application. *ACS Appl. Mater. Interfaces* **2017**, *9*, 13564–13570. [CrossRef]
20. Zhang, X.; Shi, C.; Liu, E.; Li, J.; Zhao, N.; He, C. Nitrogen-doped graphene network supported copper nanoparticles encapsulated with grapheme shells for surface-enhanced Raman scattering. *Nanoscale* **2015**, *7*, 17079–17087. [CrossRef]
21. Li, J.; Huang, Y.; Ding, Y.; Yang, Z.; Li, S.; Zhou, X.; Fan, F.; Zhang, W.; Zhou, Z.; Wu, D.; et al. Shell-isolated nanoparticle-enhanced Raman spectroscopy. *Nature* **2010**, *464*, 392–395. [CrossRef] [PubMed]
22. Yang, M.; Chen, T.; Lau, W.; Wang, Y.; Tang, Q.; Yang, Y.; Chen, H. Development of polymer–encapsulated metal nanoparticles as surface–enhanced Raman scattering probes. *Small* **2009**, *5*, 198–202. [CrossRef]
23. Shen, A.; Chen, L.; Xie, W.; Hu, J.; Zeng, A.; Richards, R.; Hu, J. Triplex Au–Ag–C core–shell nanoparticles as a novel Raman label. *Adv. Funct. Mater.* **2010**, *20*, 969–975. [CrossRef]
24. Xu, S.; Man, B.; Jiang, S.; Wang, J.; Wei, J.; Xu, S.; Liu, H.; Gao, S.; Liu, H.; Li, Z.; et al. Graphene/Cu nanoparticle hybrids fabricated by chemical vapor deposition as surface-enhanced Raman scattering substrate for label-free detection of adenosine. *ACS Appl. Mater. Interfaces* **2015**, *7*, 10977–10987. [CrossRef]
25. Huang, L.; Zhang, L.; Song, J.; Wang, X.; Liu, H. Superhydrophobic nickel-electroplated carbon fibers for versatile oil/water separation with excellent reusability and high environmental stability. *ACS Appl. Mater. Interfaces* **2020**, *12*, 24390–24402. [CrossRef]
26. Yang, W.; Li, J.; Zhou, P.; Zhu, L.; Tang, H. Superhydrophobic copper coating: Switchable wettability, on-demand oil-water separation, and antifouling. *Chem. Eng. J.* **2017**, *327*, 849–854. [CrossRef]
27. Aghajani, S.; Accardo, A.; Tichem, M. Aerosol direct writing and thermal tuning of copper nanoparticle patterns as surface-enhanced Raman scattering sensors. *ACS Appl. Nano Mater.* **2020**, *3*, 5665–5675. [CrossRef]
28. Diao, F.; Xiao, X.; Luo, B.; Sun, H.; Ding, F.; Ci, L.; Si, P. Two-step fabrication of nanoporous copper films with tunable morphology for SERS application. *Appl. Surf. Sci.* **2018**, *427*, 1271–1279. [CrossRef]
29. Yan, X.; Wang, Y.; Shi, G.; Wang, M.; Zhang, J.; Sun, X.; Xu, H. Flower-like Cu nanoislands decorated onto the cicada wing as SERS substrates for the rapid detection of crystal violet. *Optik* **2018**, *172*, 812–821. [CrossRef]
30. Lim, H.; Kim, D.; Kim, Y.; Nagaura, T.; You, J.; Kim, J.; Kim, H.; Na, J.; Henzie, J.; Yamauchi, Y. A mesopore-stimulated electromagnetic near-field: Electrochemical synthesis of mesoporous copper films by micelle self-assembly. *J. Mater. Chem. A* **2020**, *8*, 21016–21025. [CrossRef]
31. Mao, A.; Ding, M.; Jin, X.; Gu, X.; Cai, C.; Xin, C.; Zhang, T. Direct, rapid synthesis of water-dispersed copper nanoparticles and their surface-enhanced Raman scattering properties. *J. Mol. Struct.* **2015**, *1079*, 396–401. [CrossRef]
32. Ramani, T.; Prasanth, K.L.; Sreedhar, B. Air stable colloidal copper nanoparticles: Synthesis, characterization and their surface-enhanced Raman scattering properties. *Physica E* **2016**, *77*, 65–71. [CrossRef]
33. Balčytis, A.; Ryu, M.; Seniutinas, G.; Juodkazytė, J.; Cowie, B.C.C.; Stoddart, P.R.; Zamengo, M.; Morikawa, J.; Juodkazis, S. Black-CuO: Surface-enhanced Raman scattering and infrared properties. *Nanoscale* **2015**, *7*, 18299–18304. [CrossRef]
34. Proniewicz, E.; Vantasin, S.; Olszewski, T.K.; Boduszek, B.; Ozaki, Y. Biological application of water-based electrochemically synthesized CuO leaf-like arrays: SERS response modulated by the positional isomerism and interface type. *Phys. Chem. Chem. Phys.* **2017**, *19*, 31842–31855. [CrossRef]
35. Yang, L.; Liu, D.; Cui, G.; Xie, Y. Cu_2+1O/graphene nanosheets supported on three dimensional copper foam for sensitive and efficient non-enzymatic detection of glucose. *RSC Adv.* **2017**, *7*, 19312–19317. [CrossRef]
36. Cao, X.; Cui, L.; Liu, B.; Liu, Y.; Jia, D.; Yang, W.; Razal, J.M.; Liu, J. Reverse synthesis of star anise-like cobalt doped Cu-MOF/Cu_2+1O hybrid materials based on a $Cu(OH)_2$ precursor for high performance supercapacitors. *J. Mater. Chem. A* **2019**, *7*, 3815–3827. [CrossRef]
37. Wang, Y.; Liu, J.; Ozaki, Y.; Xu, Z.; Zhao, B. Effect of TiO_2 on altering direction of interfacial charge transfer in a TiO2-Ag-MPY-FePc system by SERS. *Angew. Chem. Int. Ed.* **2019**, *58*, 8172–8176. [CrossRef] [PubMed]
38. Wang, X.; Guo, L. SERS activity of semiconductors: Crystalline and amorphous nanomaterials. *Angew. Chem. Int. Ed.* **2020**, *59*, 4231–4239. [CrossRef]
39. Jensen, L.; Aikens, C.M.; Schatz, G.C. Electronic structure methods for studying surface-enhanced Raman scattering. *Chem. Soc. Rev.* **2008**, *37*, 1061–1073. [CrossRef]

MDPI

Article

Surface-Enhanced Raman Scattering Activity of ZrO_2 Nanoparticles: Effect of Tetragonal and Monoclinic Phases

Mingyue Yi [1], Yu Zhang [1], Jiawen Xu [1], Dingyuan Deng [1], Zhu Mao [2], Xiangchun Meng [1], Xiumin Shi [1,*] and Bing Zhao [3,*]

[1] College of Chemical Engineering, Changchun University of Technology, Changchun 130012, China; yimingyue95@163.com (M.Y.); zhangyjy0@163.com (Y.Z.); xujiawen20210719@163.com (J.X.); dengdingyuan0824@163.com (D.D.); mengxiangchun@ccut.edu.cn (X.M.)

[2] School of Chemistry and Life Science, Changchun University of Technology, Changchun 130012, China; maozhu@ccut.edu.cn

[3] State Kay Laboratory of Supramolecular Structure and Material, Jilin University, Changchun 130012, China

* Correspondence: shixiumin@ccut.edu.cn (X.S.); zhaob@jlu.edu.cn (B.Z.); Tel.: +86-431-85716463 (X.S.)

Abstract: The effect of the ZrO_2 crystal form on surface-enhanced Raman scattering (SERS) activity was studied. The ratio of the tetragonal (T) and monoclinic (M) phases of ZrO_2 nanoparticles (ZrO_2 NPs) was controlled by regulating the ratio of two types of additives in the hydrothermal synthesis method. The SERS intensity of 4-mercaptobenzoic acid (4–MBA) was gradually enhanced by changing the M and T phase ratio in ZrO_2 NPs. The degree of charge transfer (CT) in the enhanced 4–MBA molecule was greater than 0.5, indicating that CT was the main contributor to SERS. The intensity of SERS was strongest when the ratio of the T crystal phase in ZrO_2 was 99.7%, and the enhancement factor reached 2.21×10^4. More importantly, the proposed study indicated that the T and M phases of the ZrO_2 NPs affected the SERS enhancement. This study provides a new approach for developing high-quality SERS substrates and improving the transmission efficiency of molecular sensors.

Keywords: SERS; ZrO_2; M phase; T phase; charge transfer

Citation: Yi, M.; Zhang, Y.; Xu, J.; Deng, D.; Mao, Z.; Meng, X.; Shi, X.; Zhao, B. Surface-Enhanced Raman Scattering Activity of ZrO_2 Nanoparticles: Effect of Tetragonal and Monoclinic Phases. *Nanomaterials* **2021**, *11*, 2162. https://doi.org/10.3390/nano11092162

Academic Editors: Ronald Birke and Bing Zhao

Received: 31 July 2021
Accepted: 21 August 2021
Published: 24 August 2021

Publisher's Note: MDPI stays neutral with regard to jurisdictional claims in published maps and institutional affiliations.

1. Introduction

After the phenomenon of surface-enhanced Raman scattering (SERS) was first reported in the mid-1970s, it attracted the attention of many scholars. Researchers have published many articles related to SERS [1,2]. As a fairly new spectral detection technology, SERS has been widely used in the fields of chemistry, physics, biology, and medicine because of its high sensitivity, nondestructive detection, molecular fingerprint information, rapid and simple operation, etc. [3–5]. Single molecule SERS was studied by the Kneipp [6] and Nie [7] research groups in 1997, in which the enhancement factor (EF) reached up to 10^{14}. Nowadays, electromagnetic mechanism (EM) and chemical mechanism (CM) are widely used to explain the enhancement mechanisms of SERS [8–10]. Recently, with the evolution of SERS research, SERS substrates have changed from the original precious metals (Au, Ag and Cu) [11–13] and the expensive metals (Pt and Pd) [14,15] to semiconductor materials (ZnO, CuO, TiO_2) [16–18]. Semiconductor materials have good optics, electrics, biocompatibility, high stability, and low cost compared to the metal substrates. Therefore, they have greater research potential and application prospects [19].

A key problem in SERS research is the selection of suitable substrates and the optimization of the preparation process. Many studies have reported that zirconium dioxide (ZrO_2) has excellent properties, such as high temperature, corrosion, and oxidation resistance, thermal and chemical stability, which ensures it is widely used in the production of artificial teeth, high temperature materials, electronics and bio ceramics [20–22]. Specifically, nano-sized ZrO_2 has three types of crystal structure: monoclinic (M), tetragonal (T) and cubic

(C) [23,24], in addition to being an N-type semiconductor. Moreover, nano-ZrO_2 exhibits surface and interface, quantum size, and macroscopic quantum tunneling effects [25]. Due to the small particle size, large specific surface area and incomplete coordination of atoms on the surface, there is an increase in the quantity of active sites on the nano-ZrO_2 surface [26]. The electrons and holes, which are produced by nano-ZrO_2 crystal particles under illumination, have strong reduction and oxidation ability as well as high photoelectric conversion efficiency [27]. The application scope of semiconductor materials is expanding with the development of the semiconductor technique. The use of semiconductor materials as SERS substrates is prevalent; however, to date, only two studies describing the use of ZrO_2 as a SERS substrate have been reported. The crystalline form of nano-ZrO_2 is mainly the M phase [24,28] and the EF can reach 10^3. Nano-ZrO_2 is very suitable as a SERS substrate due to its superior characteristics, and using nano-ZrO_2 expands the range of semiconductor nanomaterials as SERS substrates. Meanwhile, it also leads to new applications of nano-ZrO_2, which lay a foundation for the further research of the high-quality material in the application of nano sensor devices. More suitable preparation techniques of nano-ZrO_2 are developed by improving the preparation process. The internal relationship between the SERS properties of a substrate and its structure is a contentious issue in SERS research.

In this work, nano-ZrO_2 was prepared by hydrothermal synthesis and the SERS of 4-mercaptobenzoic acid (4–MBA) adsorbed on ZrO_2 nanoparticles (NPs) was determined. The different crystal forms of ZrO_2 NPs can be prepared by adding different types of additives in the hydrothermal synthesis. More interestingly, the proportion of the different crystal forms of ZrO_2 NPs can be controlled by regulating the proportion of two types of additives. We found that there is an interdependence between the crystal form of ZrO_2 NPs and the intensity of the SERS signal.

2. Materials and Methods

2.1. Chemicals

Zirconium chloride octahydrate ($ZrOCl_2 \cdot 8H_2O$), 1,2-dichloroethane, and diethanolamine were purchased from Shanghai McLean Biochemistry Co., LTD. (Shanghai, China). Ammonia water (25%) and 4–MBA were purchased from Tianjin Furui Fine Chemical Co., LTD (Tianjin, China) and Sigma-Aldrich International Limited (Shanghai, China). Deionized water was used throughout the experiments. All reagents used were analytical grade and were used as supplied for sample preparation.

2.2. Preparation of ZrO_2 NPs

ZrO_2 NPs were prepared using the hydrothermal synthesis method according to the literature [29]. $ZrOCl_2 \cdot 8H_2O$ (3.9710 g) was added to 12 mL of deionized water and dissolved completely with stirring. Then, ammonia solution (10%) was added to the solution until the pH was approximately 7, and a white precipitate was produced. The precipitate was filtered repeatedly and washed with deionized water. The precipitate was transformed into a homogeneous suspension in deionized water, and the pH was adjusted to 9 with the addition of ammonia solution (5 wt %). The concentration of zirconium ions was then adjusted to 0.6 mol/L. Diethanolamine and 1,2-dichloroethane were utilized as a mixed additive to form the different crystal structures of ZrO_2 NPs. The mixing ratios were 1:0, 4:1, 3:1, 2:1, 1:1, 1:2, 1:3, 1:4 and 0:1, respectively. One of the nine ratios of mixed additives was added to each sample of the suspension. The volume ratio of the additives to the suspension was 1:30. The volume of the suspension was adjusted to 16 mL using deionized water. The reaction solution was then stirred for 40 min and placed in a high-pressure hydrothermal reaction vessel. Nine groups of reactors were heated at 180 °C for 72 h. After the reaction, the synthesized ZrO_2 was filtered, washed several times, and dried at room temperature.

2.3. *Adsorption of Probe Molecules for SERS Measurement*

ZrO_2 (30 mg) with different T phase ratios was dispersed in 15 mL 4-MBA (1×10^{-3} mol/L) ethanol solution. The configured ZrO_2 suspension was magnetically stirred for 8 h at room temperature. After centrifugation, ethanol and deionized water were used to wash the residue trice, and the surface modification of ZrO_2 NPs was obtained.

2.4. *Instrumentation*

The crystal structure of all the samples was determined using Rikagu Smartlab X-ray diffraction patterns (Tokyo, Japan), with a scanning angle range of 10°–75°, and a scanning speed of 3°/min. SERS spectra were recorded by a Horiba HR Evolution Raman spectrometer (Paris, France) under 532 nm (or 2.33 eV) excitation. All Raman spectra were calibrated using Si plates. The penetration depth and diameter of the laser spot was approximately 10 and 1.3 μm, respectively. Scanning electron microscopy (SEM) was employed to study the size distribution of ZrO_2 NPs using a JEOL-7610P (Tokyo, Japan). Ultraviolet-visible (UV-Vis) diffuse reflectance spectra (DRS) of the ZrO_2 samples were recorded on a Cary 5000 UV-Vis spectrophotometer (Santa Clara, CA, USA) equipped with an integrating sphere using $BaSO_4$ as a reference.

3. Results and Discussion

3.1. *Characterization of ZrO_2 NPs*

Previous reports [29,30] describe the preparation of ZrO_2 NPs with different crystal forms using different additives during hydrothermal synthesis. Diethanolamine with its strong chelating ability encourages the formation of T–ZrO_2 crystals, whereas alkyl halide 1,2-dichloroethane encourages the formation of M–ZrO_2 crystals. Figure 1 shows the XRD diagram of ZrO_2 NPs prepared by a hydrothermal method using diethanolamine and 1,2-dichloroethane as a mixed additive, and the mixing ratios are 1:0; 4:1; 3:1; 2:1; 1:1; 1:2, 1:3, 1:4, and 0:1, respectively. As shown in Figure 1, there are obvious differences in the crystal form of nano-ZrO_2 prepared with pure diethanolamine or 1,2-dichloroethane as dispersant. The XRD diffraction curve when using diethanolamine as an additive is shown in Figure 1a, the characteristic Bragg reflections for the T phase: T (101), T (200), T (202) and T (131) (PDF card: 50–1089) are respectively at approximately 2θ = 30.45°, 35.30°, 50.26° and 60.2°. This indicates that the phase of ZrO_2 is basically pure T phase [31]. However, the XRD diffraction curve when using 1,2-dichloroethane as an additive is shown in Figure 1i, the characteristic Bragg reflections for the M phase: M (111) and M (−111) (PDF card: 37-1484) are at approximately 2θ = 28.37° and 31.50°, respectively. This indicates that the ZrO_2 crystal phase is basically pure M. Figure 1b–h are the XRD of ZrO_2 NPs prepared using the diethanolamine and 1,2-dichloroethane mixture, in the mixing ratios stipulated above. We concluded that the intensity of the Bragg reflection of the T phase characteristic in the diffraction line (Figure 1b–h) decreases gradually, whereas the Bragg reflection of the M phase characteristic increases gradually with the decrease diethanolamine and the increase 1,2-dichloroethane. The particle size of ZrO_2 NPs was calculated using the Scherrer formula: $D = k\lambda/(\beta \cos \theta)$ [32] and is approximately 15.5–16.7 nm. The particle size of the ZrO_2 NPs was confirmed using the SEM image (Figure S1 and Table S1) and was consistent with the results from the XRD.

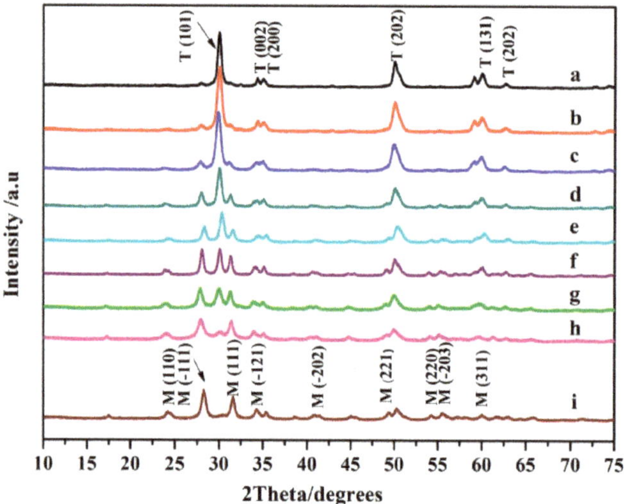

Figure 1. XRD patterns of ZrO_2 NPs prepared with different ratios of diethanolamine and 1,2-dichloroethane as mixed additives. (**a–i**) correspond to the mixing ratio of 1:0, 4:1, 3:1, 2:1, 1:1, 1:2, 1:3, 1:4, and 0:1, respectively.

It is necessary to calculate the T phase content in the sample ZrO_2. To determine the correlation between the mixing ratio of diethanolamine and 1,2-dichloroethane additives, and the crystal form of the synthesized ZrO_2 NPs. The T phase and M phase of ZrO_2 can be identified by the most characteristic Bragg reflection of the XRD diffraction lines of all the samples: M (−111), T (101) and M (111). There were two M peaks (−111) and (111), and one T peak (101) in the sample. The T phase content in ZrO_2 was calculated by the relative intensities. First, the strength of the integral of the three Bragg reflections: M (−111), T (101), and M (111) of the samples were extracted. Then we calculated the ratio of the T phase using the empirical expressions of Garvie and Nicholson [33,34]. The Formulas (1) and (2) are as follows:

$$X_t = \frac{I_t(101)}{[I_t(101) + I_m(-111) + I_m(111)]} \tag{1}$$

$$v_t = \frac{1.311X_t}{(1 + 0.311X_t)} \tag{2}$$

where v_t is the volume fraction of the T phase, X_t is the integrated intensity ratio; I_m (111) and I_m (−111) are the (111) and (−111) intensities of the M phase of ZrO_2, and I_t (101) is the (101) intensity of the T phase of ZrO_2, respectively.

The T phase content of the ZrO_2 crystal in Figure 1a–i were calculated by Formulas (1) and (2). The ratio of the T phase in the ZrO_2 nanoparticles are 99.7%, 81.3%, 70.6%, 63.5%, 41.1%, 29.8%, 19.4%, 11.3%, and 2.5% corresponding to Figure 1a–i, respectively. The results show that ZrO_2 with different crystal forms can be synthesized by different types of organic additives in the process of hydrothermal synthesis. The T phase content of ZrO_2 NPs can be easily controlled by adjusting the ratio of diethanolamine to 1,2-dichloroethane. The content of the T phase of ZrO_2 NPs gradually decreased, with decreasing diethanolamine and increasing 1,2-dichloroethane (Figure 1).

3.2. Raman Spectra of ZrO_2 NPs

The Raman spectra of the synthesized ZrO_2 NPs are shown in Figure 2. In general, Raman spectra of molecules appear due to the vibrational characteristics of different crystal phases of ZrO_2. The Raman shifts at less than 800 cm^{-1} were assigned to the ZrO_2 NPs. There were a few distinct vibrational bands located at 145, 266, 313, 458 and 646 cm^{-1} which correspond to the vibrational modes of the T phase of ZrO_2 (Figure 2a). For the T phase of ZrO_2, 6 modes ($1A_{1g} + 3E_g + 2B_{1g}$) are Raman active [35]; the E representations are two dimensional whereas the other modes are one dimensional. This suggests that the ZrO_2 is essentially pure T phase when diethanolamine is used as an additive. However, there were a few distinct vibrational bands located at 177, 188, 328, 338, 378, 475, and 612 cm^{-1} which correspond to the vibrational modes of the M phase of ZrO_2 (Figure 2i). For the M phase of ZrO_2, 18 modes ($9A_g + 9B_g$) are Raman active [36]. This indicates that the ZrO_2 is essentially pure M phase when 1,2-dichloroethane is used as an additive. The curves (b–h) are the Raman spectra of ZrO_2 NPs prepared using diethanolamine and 1,2-dichloroethane as additives, in the stipulated ratios. As shown in Figure 2b–h, the vibrational mode intensity of the T phase of ZrO_2 gradually decreases, whereas the vibration mode of M phase of ZrO_2 gradually increases, with decreasing diethanolamine and increasing 1,2-dichloroethane. This indicates that the results of the Raman and XRD analyses are consistent.

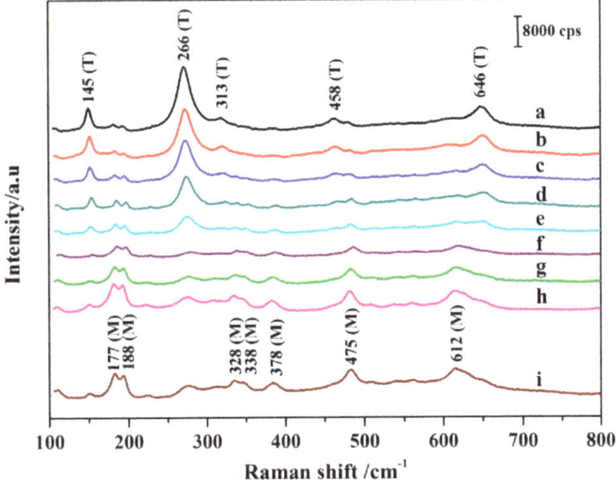

Figure 2. Raman spectra of ZrO_2 NPs prepared with different ratios of diethanolamine and 1,2-dichloroethane as mixed additives. (**a–i**) correspond to the mixing ratio: 1:0, 4:1, 3:1, 2:1, 1:1, 1:2, 1:3, 1:4, and 0:1, respectively.

According to a previous report [37], the proportion of specific crystal forms of molecules can be estimated by using the characteristic bands of different crystal phases in the Raman spectra of molecules. As is illustrated in Figure 2 (curves a and i), the T and M phases of ZrO_2 can be identified by the most representative Raman displacement located at the following positions: T (145, 266, and 646 cm^{-1}) and M (177, 188, and 612 cm^{-1}). The T phase content in ZrO_2 was calculated by the relative intensities of the T bands located at 145, 266, and 646 cm^{-1}, and the M bands located at 177, 188 and 612 cm^{-1}. The ratios of the T phase of ZrO_2 were calculated using the empirical expressions of Clarke and Adar [38]. The formula is as follows (3):

$$\nu_t = \frac{0.97[I_t(145) + I_t(266) + I_t(646)]}{0.97[I_t(145 + I_t(266) + It(646)] + [I_m(177) + I_m(188) + I_m(612)]} \quad (3)$$

where m and t represent the M and T phases, ν_t is the ratio of the T phase of ZrO_2, I_m is the intensity of the M phase of ZrO_2, and I_t is the intensity of the T phase of ZrO_2, respectively. The ratio of the T phase in ZrO_2 NPs is: 99.6%, 82.0%, 70.1%, 63.4%, 41.5%, 29.7%, 19.0%,

11.5%, and 2.6% corresponding to a–i in Figure 2, respectively. These calculation results are in good agreement with those calculated using the characteristic diffraction peaks of XRD.

3.3. UV-Vis DRS Spectra of ZrO_2 NPs

The optical properties of ZrO_2 NPs were studied by the UV-Vis DRS technique. Figure 3 shows the absorbance spectra of ZrO_2 at different crystal ratios. The UV-Vis absorption spectrum of the semiconductor can determine its band gap energy and surface defect state [39]. The strong absorption band of nine samples at less than 250 nm is due to the band-band energy transition of ZrO_2 NPs. The trailing absorption bands at 250–430 nm are assigned to the surface defect state of ZrO_2. The optical band gap and incident photon energy correspond to the transformed Kubelka–Munk function [40]. The calculated value of the band gap energy (Eg(eV))and the photo-absorption thresholds (λ_g(nm)) can be achieved from the Figure S2 and Table S2 of the Supplementary Information. Thus, the Eg with λ_g of ZrO_2 which corresponds to the curves of a–i are 2.96 (419), 3.16 (392), 3.41 (364), 3.66 (339), 3.81 (325), 3.95 (314), 4.22 (294), 4.41 (281), and 4.68 eV (265 nm), respectively [41]. It is noteworthy that with the increase of the T phase ratio in ZrO_2, the λ_g of band-band Transition for ZrO_2 are red shifted, and the intensity of the tail absorption band at 250–430 nm is gradually enhanced. The greater the red shift of the optical absorption threshold, the higher the threshold, the smaller its band gap energy. The greater the strength of the trailing absorption bands indicates the stronger the surface defect strength. When the ratio of the T crystal phase in ZrO_2 is the greatest, the absorbance bands assigned to the ZrO_2 have the largest red shift and the strongest strength of the trailing absorption band, thus indicating that it has the best abundant surface properties. The smaller the band gap energy and the better abundant surface defect state of the semiconductor, the more favorable the electron transfer.

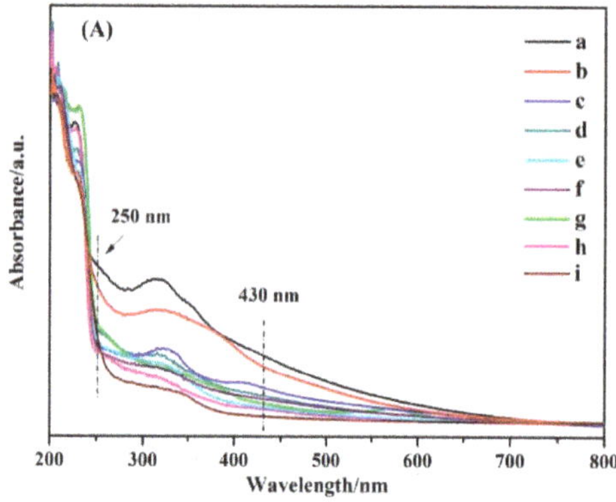

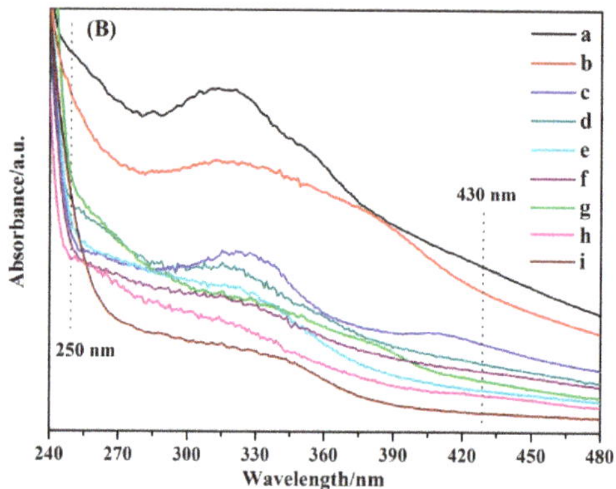

Figure 3. (**A**) UV-Vis DRS spectra of ZrO_2 NPs with different T phase proportions; (**B**) UV-Vis DRS spectra of ZrO_2 NPs with different T phase proportions in the range 240–480 nm.

3.4. Crystal Form-Dependence of SERS

Generally, ZrO_2 have M, T, and C crystal forms, and the different crystal forms have different grain boundary structures. However, the structure of the grain boundary will affect the electrical and optical properties of ZrO_2. To examine the effect of the crystal form of ZrO_2 on the SERS, Figure 4A shows the SERS spectra of 4–MBA adsorbed on ZrO_2 NPs with different T phase proportions. Notably, as the T phase ratio increases and the M phase ratio decreases, the SERS signal of 4–MBA adsorbed on the ZrO_2 NPs gradually increases. When the ratio of the T phase in ZrO_2 NPs is close to that of the pure T phase, the SERS signal is at maximum. When the ratio of the M phase in ZrO_2 NPs is close to that of the pure M phase, the SERS intensity decreases significantly. The strongest Raman intensity was obtained from the pure T phase, which is approximately four times that of the pure M phase (Figure 4B). Thus, the interdependency between the form of the crystalline phase in ZrO_2 NPs and the SERS enhancement effect can be confirmed. The T phase of ZrO_2 favors CT in the SERS enhancement more than the M phase. This is because of the effect of the grain boundary structure of the different crystalline phases on the electrical and optical properties of ZrO_2. In addition, the T crystalline phase of ZrO_2 has more abundant surface states, which is more conducive to the occurrence of CT.

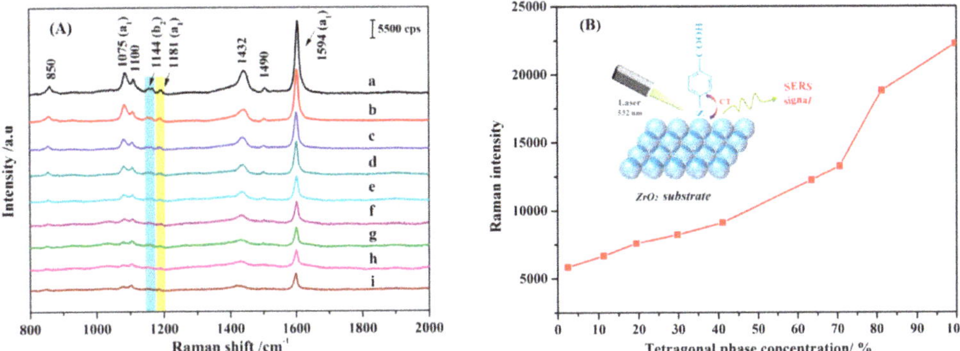

Figure 4. (**A**) SERS spectra of 4–MBA adsorbed on ZrO_2 NPs with different T phase proportions. (**B**) The relationship between SERS intensities of 4–MBA at the 1594 cm^{-1} mode and the T phase concentration in ZrO_2 NPs.

3.5. Evaluation of EF

To evaluate the SERS activity of ZrO_2 NPs, EF was employed to reflect the enhancement ability. To accurately calculate the values of EF, the Raman spectra of 4–MBA adsorbed on ZrO_2 NPs, and of the solid 4–MBA powders were obtained as shown in Figure 5. The calculation method adopts Formula (4) [42]:

$$EF = \frac{I_{Surf}}{I_{Bulk}} \times \frac{N_{Bulk}}{N_{Surf}} \tag{4}$$

where I_{surf} and I_{Bulk} represent the SERS intensity of 4–MBA adsorbed on the ZrO_2 NPs and 4–MBA solid powders at 1594 cm^{-1} (assigned to the characteristic vibration of the ν(C–C) aromatic ring), respectively. N_{Surf} and N_{Bulk} represent the number of molecules adsorbed on ZrO_2 NPs and 4–MBA solid powders, respectively.

The value of I_{surf}/I_{Bulk} is equal to the Raman intensity ratio of the 4–MBA@ZrO_2 to the bulk 4–MBA in Figure 5, and the calculated result is 1.8. N_{Bulk} (the number of bulk 4–MBA detected in the 1.3 μm laser spot) was estimated to be 7.61×10^{10} based on 532 nm laser excitation, the focusing penetration depth of the laser (~10 μm), and the density of the 4–MBA (1.5 g·cm^{-3}) [43]. If the ZrO_2 NPs are assumed to be uniformly distributed in a single layer, and the boundary density of 4–MBA adsorbed on the NPs surface is 0.5 nmol/cm^2, the calculated N_{Surf} is 6.146×10^6 [43]. The results of the above calculation are substituted into Formula (4), and the value of EF is 2.21×10^4.

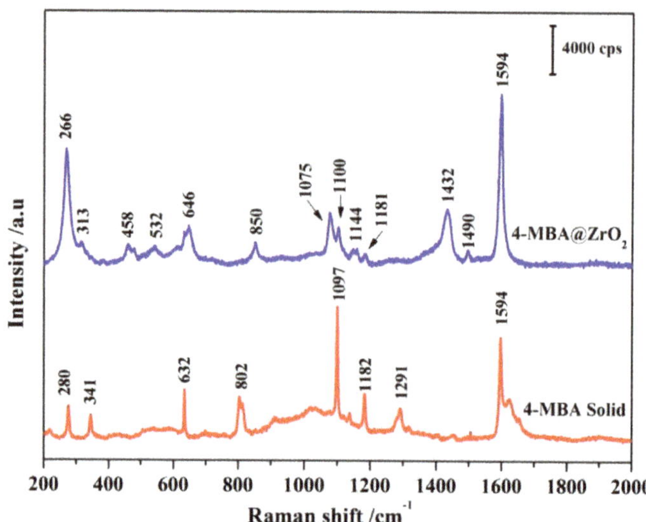

Figure 5. SERS spectra of 4–MBA adsorbed on ZrO_2 NPs and Raman spectra of the 4–MBA bulk sample.

3.6. Degree of Charge Transfer

CT is also the major contribution to SERS in addition to surface plasmon resonance (SPR) and molecular transition [44]. In accordance with previous reports, we employed the degree of CT ($\varrho_{CT}(k)$) to measure the CT contribution to SERS. The concept of $\varrho_{CT}(k)$ can be defined by the following Formula (5) [45]:

$$\varrho CT(k) = \frac{I^k(CT) - I^k(SPR)}{I^k(CT) + I^0(SPR)} \quad (5)$$

where k refers to a specific band in the Raman spectrum. In this equation, $I^k(CT)$ and $I^k(SPR)$ are the intensities of the k band in the presence of CT and EM contribution, respectively, whereas $I^0(SPR)$ is the intensity of the fully symmetric vibrational mode, selected in the spectral region with only SPR contribution. k can be a fully symmetric or non-fully symmetric line. For a fully symmetric line, $I^k(SPR) = I^0(SPR)$, but in the latter, $I^k(SPR)$ is usually quite small [46,47], and we can think of it as being negligible, that is, I^k (SPR) = 0. It can be seen from Formula (5) that when $\varrho_{CT} = 0$, there is no CT contribution, and when $\varrho_{CT} = 1$, the intensity of SERS mainly comes from the CT effect. When $\varrho_{CT} = 1/2$, the contribution of CT and EM is equal. As shown from Figure 6A, because the peak at 1144 cm^{-1} is an independent b_2 vibrational mode, it is selected as the research object of CT contribution, that is, I^k (CT). The frequency band at 1181 cm^{-1} is an independent a_1 vibrational mode, which is selected as I^0 (SPR). After selecting the contribution peaks of the corresponding objects, the relationship of ϱ_{CT} and the ZrO_2 NPs with different T phases were shown in Figure 6B. The ϱ_{CT} value indicates that when the ratio of diethanolamine to 1,2-dichloroethane is 1:0, and the purity of T zirconia is 99.7%, the CT degree is the strongest. With a decrease in purity, the CT degree gradually decreases, and all of them are greater than 0.5, indicating that CT is the main contributor to SERS intensity.

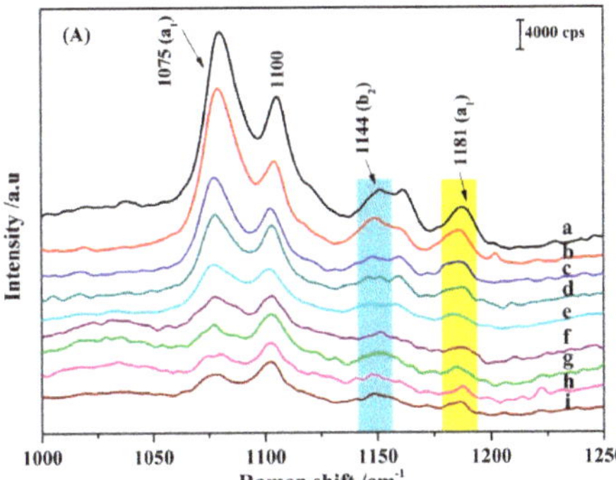

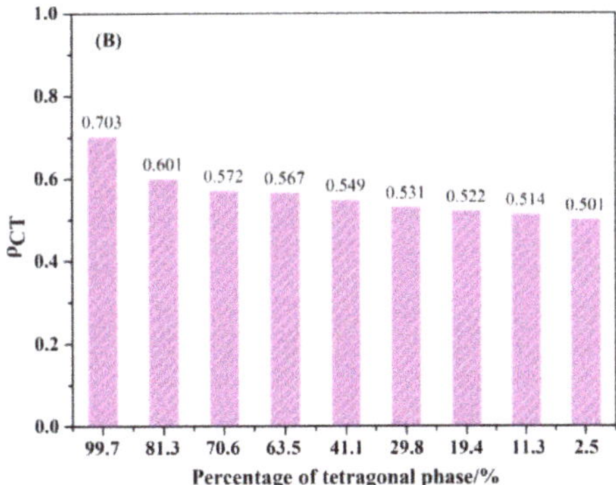

Figure 6. (**A**) SERS spectra of 4–MBA adsorbed on ZrO_2 NPs with different T phase proportions. (**B**) Degree of charge transfer (ϱ_{CT}) values of the different T phase proportions of ZrO_2 NPs based on 532 nm laser excitation.

3.7. SERS Mechanisms of ZrO_2 NPs

The enhancement mechanism of SERS has always been a core contentious issue in the research of SERS substrate development. The study of the SERS enhancement mechanism provides theoretical guidance for the development of high-quality SERS substrates. There are two widely accepted SERS mechanisms to expound the SERS influence. One is EM and the other is CM. The EM effect is due to the local electric field generated by the collective oscillation of the plasma on the metal surface [48,49]. The CM effect is because the CT process between the substrate and the molecule can induce CM. The CT processes typically occur between metal or semiconductor substrates, and the probe molecules. As the SPR is some distance away from the visible region, the SERS effect of most semiconductor materials is due to the CT between semiconductor materials and molecules. As shown in 4–MBA@ZrO_2 of Figure 5, the Raman spectra of 4–MBA on ZrO_2 are similar to those of 4–MBA on other semiconductor materials reported in the literature. Two stronger Raman bands at 1075 and 1594 cm^{-1} are attributed to the ν_{12} (a_1) and ν_{8a} (a_1) modes of the respiratory vibration in the aromatic ring, respectively. The other two weaker Raman bands at 1144 and 1181 cm^{-1} are assigned to the ν_{15} (b_2) and ν_9 (a_1) [50,51] modes of the bending deformation vibration, respectively.

In contrast, it can be seen from the comparison of the UV absorption spectra of ZrO_2 and 4–MBA@ZrO_2 (Figure 7), that after 4–MBA adsorption, the intensity of the band energy transition absorption band of ZrO_2, at less than 250 nm, has increased, and the red shift is significant. At 250–430 nm, the increase in the tail band intensity was greater and the degree of the red shift was larger than that at 250 nm. These changes indicate a CT process between the 4–MBA and the ZrO_2 NPs [52]. The obvious changes in the tail band intensity and the degree of the red shift of the bands at 250–430 nm indicated that the surface state energy level, caused by surface defects, plays an important role in the CT process. The degree of surface defects directly affects the CT process and eventually leads to the SERS effect. Similar results were found in the Cu_2O–4–MBA system [53].

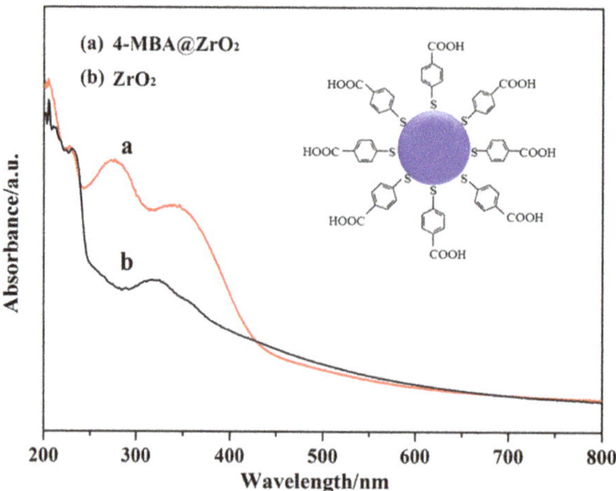

Figure 7. UV-Vis DRS spectra of (**a**) 4–MBA adsorbed on the ZrO_2 NPs and (**b**) ZrO_2 NPs (the ratio of T phase is 99.7%). The inset is the schematic diagram of 4–MBA adsorbed on ZrO_2 NPs.

The CT process in the SERS system of a semiconductor-molecule depends on the vibration coupling between the semiconductor and molecular energy levels. The direction of CT relies on the energy levels of the valence (VB) and conduction bands (CB) of the semiconductor, relative to the highest occupied molecular orbitals (HOMO), and the lowest unoccupied molecular orbitals (LUMO) of the adsorbed molecules. The HOMO and LUMO of the adsorption molecule 4–MBA were −8.48 and −3.85 eV, based on previous literature [54,55]. The VB of ZrO_2 is located at −7.6 eV [56]. As the band gap energy of ZrO_2 is 2.96 eV, the calculated CB of ZrO_2 is located at −4.64 eV, (Figure 8). As ZrO_2 with a large amount of oxygen vacancies enriching the surface states, there are transition energy levels (surface energy levels) between the VB and CB of ZrO_2. A possible CT transfer pattern is shown in Figure 8; incident light of 532 nm (2.33 eV) excites electrons from the VB of the ZrO_2 through the surface state (Ess) energy level to the CB of ZrO_2, then transits to the LUMO of the 4–MBA, and finally radiates out the Raman photons. The surface defects can be explained by using the UV-Vis DRS spectra. It can be seen from the curve i–a in the Figure 3 that with the increase of the T phase ratio in ZrO_2, the intensity of the tail absorption band at 250–430 nm is gradually enhanced, indicating that the degree of surface defect state gradually increases. The rich surface defect state is more beneficial to the SERS effect during CT transfer. With the increase of the T crystal form ratio, more surface defects are produced, which lead to the stronger SERS effect. The experimental results are consistent with the above enhancement mechanism, which indicates that the SERS effect of the ZrO_2 substrate can be improved by adjusting the ratio of the T crystalline phase in the ZrO_2.

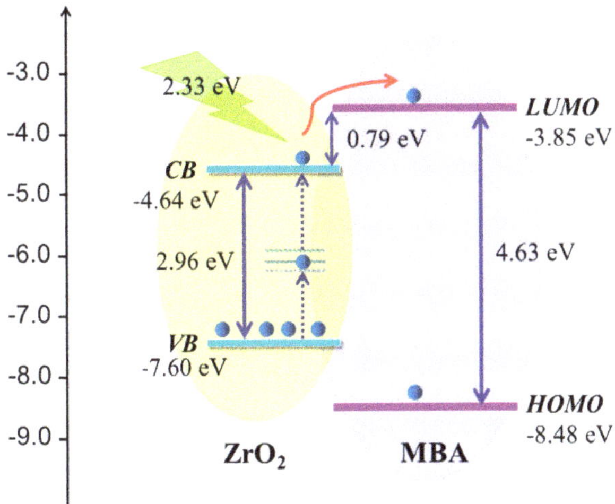

Figure 8. Diagram of the CT mechanism of 4–MBA–ZrO_2 system.

4. Conclusions

ZrO_2 NPs with different ratios of the crystalline phase were synthesized by hydrothermal synthesis and adjusting the ratio of two additives. The additive, diethanolamine, encourages the formation of T–ZrO_2 crystals, whereas 1,2-dichloroethane encourages the formation of M–ZrO_2 crystals. The ratio of the T and M phases of ZrO_2 NPs can be easily controlled. When the probe molecules of 4–MBA are adsorbed onto the ZrO_2, the SERS intensity of 4–MBA is gradually enhanced with the increasing proportion of the T phase of ZrO_2 NPs. The strongest SERS intensity of 4–MBA can be obtained by using the pure T phase. The value of EF reached 2.21×10^4. The UV-Vis DRS spectra have shown the interrelation between the band gap energy, surface defect state and the proportion of T phase in ZrO_2 NPs. As the proportion of T phase in ZrO_2 NPs increases, the band gap energy decreases but the degree of surface defects increases gradually. The smaller band gap energy and the richer surface defect states are conducive to CT transfer between ZrO_2 NPs and 4–MBA, consequently showing much stronger SERS signal. This is the main reason that the SERS effect of the T–ZrO_2 substrate is stronger than that of the M–ZrO_2 substrate. The schematic diagram of CT shows the direction of CT and further indicates that the crystal form of ZrO_2 NPs is an important factor affecting the CT. The SERS effect of semiconductor substrates was examined from the perspective of the crystal form of semiconductor nanomaterials in an unprecedented attempt. This study illustrates a new way to develop better quality semiconductor substrates and highlights a new method to improve the charge transport rate for the construction of high sensitivity molecular sensors.

Supplementary Materials: The following are available online at https://www.mdpi.com/article/10.3390/nano11092162/s1, Figure S1: SEM images of ZrO_2 NPs prepared with different ratios of diethanolamine and 1,2-dichloroethane as mixed additives. a–i correspond to mixing ratio of diethanolamine and 1,2-dichloroethane are 1:0, 4:1, 3:1, 2:1, 1:1, 1:2, 1:3, 1:4, and 0:1, respectively, Figure S2: UV-Vis DRS spectra of ZrO_2 NPs prepared with different ratios of diethanolamine and 1,2-dichloroethane as mixed additives. a–i correspond to mixing ratio of diethanolamine and 1,2-dichloroethane are 1:0, 4:1, 3:1, 2:1, 1:1, 1:2, 1:3, 1:4, and 0:1, respectively.

Author Contributions: Writing—original draft preparation, M.Y. and Y.Z.; data curation: J.X. and D.D.; methodology, Z.M. and X.M.; writing—review and editing, X.S. and B.Z. All authors have read and agreed to the published version of the manuscript.

Funding: This research was funded by Jilin Province Science and Technology Research Project (No. 20150204024GX). This work was also supported by the National Science Foundation of China (No.21503021 and No.51673030) and the Natural Science Foundation of Jilin Province Project (No. 20200201095JC).

Institutional Review Board Statement: Not applicable.

Informed Consent Statement: Not applicable.

Data Availability Statement: Do not have the supporting reported results.

Conflicts of Interest: The authors declare no conflict of interest.

References

1. Zhang, C.; You, T.; Yang, N.; Gao, Y.; Jiang, L.; Yin, P. Hydrophobic paper-based SERS platform for direct-droplet quantitative determination of melamine. *Food Chem.* **2019**, *287*, 363–368. [CrossRef]
2. Xu, L.J.; Lei, Z.C.; Li, J.; Zong, C.; Yang, C.J.; Ren, B. Label-free Surface-enhanced Raman Spectroscopy Detection of DNA with Single-base Sensitivity. *J. Am. Chem. Soc.* **2015**, *137*, 5149–5154. [CrossRef] [PubMed]
3. Zhang, C.; You, E.; Jin, Q.; Yuan, Y.; Xu, M.; Ding, S.; Yao, J.; Tian, Z. Observing the dynamic "hot spots" on two-dimensional Au nanoparticles monolayer film. *Chem. Commun. (Camb.)* **2017**, *53*, 6788–6791. [CrossRef]
4. Chen, L.; Sa, Y.; Park, Y.; Hwang, H.; Ji, H.; Zhao, B.; Jung, Y.M. Au-MPY/DTNB@SiO_2 SERS Nanoprobe for Immunosorbent Assay. *Vib. Spectrosc.* **2016**, *87*, 34–39. [CrossRef]
5. Ji, W.; Song, W.; Tanabe, I.; Wang, Y.; Zhao, B.; Ozaki, Y. Semiconductor-enhanced Raman scattering for highly robust SERS sensing: The case of phosphate analysis. *Chem. Commun. (Camb.)* **2015**, *51*, 7641–7644. [CrossRef] [PubMed]
6. Emory, S.R.; Nie, S. Probing single molecules and single nanoparticles by surface-enhanced raman scattering. *Diss. Indiana Univ.* **1999**, *275*, 1102–1106.
7. Doering, W.E.; Nie, S. Single-Molecule and Single-Nanoparticle SERS: Examining the Roles of Surface Active Sites and Chemical Enhancement. *J. Phys. Chem. B* **2002**, *106*, 311–317. [CrossRef]
8. Kozhina, E.P.; Andreev, S.N.; Tarakanov, V.P.; Bedin, S.A.; Doludenko, I.M.; Naumov, A.V. Study of Local Fields of Dendrite Nanostructures in Hot Spots Formed on SERS-Active Substrates Produced via Template-Assisted Synthesis. *Bull. Russ. Acad. Sci. Phys.* **2020**, *84*, 1465–1468. [CrossRef]
9. Kozhina, E.P.; Bedin, S.A.; Nechaeva, N.L.; Podoynitsyn, S.N.; Tarakanov, V.P.; Andreev, S.N.; Grigoriev, Y.V.; Naumov, A.V. Ag-Nanowire Bundles with Gap Hot Spots Synthesized in Track-Etched Membranes as Effective SERS-Substrates. *Appl. Sci.* **2021**, *11*, 1375. [CrossRef]
10. Fateixa, S.; Nogueira, H.I.; Trindade, T. Hybrid nanostructures for SERS: Materials development and chemical detection. *Phys. Chem. Chem. Phys.* **2015**, *17*, 21046–21071. [CrossRef] [PubMed]
11. Ma, X.; Turasan, H.; Jia, F.; Seo, S.; Wang, Z.; Liu, G.L.; Kokini, J.L. A novel biodegradable ESERS (enhanced SERS) platform with deposition of Au, Ag and Au/Ag nanoparticles on gold coated zein nanophotonic structures for the detection of food analytes. *Vib. Spectrosc.* **2020**, *106*, 103013. [CrossRef]
12. El-Aal, M.A.; Seto, T.; Matsuki, A. The effects of operating parameters on the morphology, and the SERS of Cu NPs prepared by spark discharge deposition. *Appl. Phys. A* **2020**, *126*, 572. [CrossRef]
13. Yu, Z.; Park, Y.; Chen, L.; Zhao, B.; Jung, Y.M.; Cong, Q. Preparation of a Superhydrophobic and Peroxidase-like Activity Array Chip for H_2O_2 Sensing by Surface-Enhanced Raman Scattering. *ACS Appl. Mater. Interfaces* **2015**, *7*, 23472–23480. [CrossRef] [PubMed]
14. Shin, J.H.; Kim, H.G.; Baek, G.M.; Kim, R.; Jeon, S.; Mun, J.H.; Lee, H.-B.-R.; Jung, Y.S.; Kim, S.O.; Kim, K.N.; et al. Fabrication of 50 nm scale Pt nanostructures by block copolymer (BCP) and its characteristics of surface-enhanced Raman scattering (SERS). *RSC Adv.* **2016**, *6*, 70756–70762. [CrossRef]
15. Kundu, S.; Yi, S.I.; Ma, L.; Chen, Y.; Dai, W.; Sinyukov, A.M.; Liang, H. Morphology Dependent Catalysis and Surface Enhanced Raman Scattering (SERS) Studies using Pd Nanostructures in DNA, CTAB and PVA Scaffolds. *Dalton Trans.* **2017**, *46*, 9678–9691. [CrossRef] [PubMed]
16. Lin, J.; Yu, J.; Akakuru, O.U.; Wang, X.; Yuan, B.; Chen, T.; Guo, L.; Wu, A. Low Temperature-boosted High Efficiency Photo-induced Charge Transfer for Remarkable SERS Activity of ZnO nanosheets. *Chem. Sci.* **2020**, *11*, 9414–9420. [CrossRef] [PubMed]
17. Martinez Nuñez, C.E.; Delgado-Beleño, Y.; Rocha-Rocha, O.; Calderón-Ayala, G.; Florez-López, N.S.; Cortez Valadez, M. Chemical bonding mechanism in SERS effect of pyridine by CuO nanoparticles. *J. Raman Spectrosc.* **2019**, *50*, 1395–1404. [CrossRef]
18. Tanabe, I.; Ozaki, Y. Consistent changes in electronic states and photocatalytic activities of metal (Au, Pd, Pt)-modified TiO_2 studied by far-ultraviolet spectroscopy. *Chem. Commun. (Camb.)* **2014**, *50*, 2117–2119. [CrossRef]
19. Han, X.X.; Ji, W.; Zhao, B.; Ozaki, Y. Semiconductor-Enhanced Raman Scattering: Active Nanomaterials and Applications. *Nanoscale* **2017**, *9*, 4847–4861. [CrossRef] [PubMed]
20. Zhao, N.; Pan, D.; Nie, W.; Ji, X. Two-Phase Synthesis of Shape-Controlled Colloidal Zirconia Nanocrystals and Their Characterization. *J. Am. Ceram. Soc.* **2006**, *128*, 10118. [CrossRef] [PubMed]

21. Aati, S.; Akram, Z.; Ngo, H.; Fawzy, A.S. Development of 3D printed resin reinforced with modified ZrO_2 nanoparticles for long-term provisional dental restorations. *Dent. Mater.* **2021**, *37*, e360–e374. [CrossRef]
22. Cao, W.; Kang, J.; Fan, G.; Yang, L.; Li, F. Fabrication of Porous ZrO_2 Nanostructures with Controlled Crystalline Phases and Structures via a Facile and Cost-Effective Hydrothermal Approach. *Ind. Eng. Chem. Res.* **2015**, *54*, 12795–12804. [CrossRef]
23. Gupta, T.K.; Bechtold, J.H.; Kuznicki, R.C.; Cadoff, L.H.; Rossing, B.R. Stabilization of tetragonal phase in polycrystalline zirconia. *J. Mater. Sci.* **1997**, *12*, 2421–2426. [CrossRef]
24. Ji, P.; Mao, Z.; Wang, Z.; Xue, X.; Zhang, Y.; Lv, J.; Shi, X. Improved Surface-Enhanced Raman Scattering Properties of ZrO_2 Nanoparticles by Zn Doping. *Nanomaterials* **2019**, *9*, 983. [CrossRef] [PubMed]
25. Peymani, R.; Poursalehi, R.; Yourdkhani, A. DC Arc discharge synthesized zirconia nanoparticles: Shed light on arc current effects on size, crystal structure, optical properties and formation mechanism. *Mater. Res. Express* **2019**, *6*, 112276. [CrossRef]
26. Liu, Q.; Jiang, L.; Guo, L. Precursor-directed self-assembly of porous ZnO nanosheets as high-performance surface-enhanced Raman scattering substrate. *Small* **2014**, *10*, 48–51. [CrossRef] [PubMed]
27. Ezhil Vilian, A.T.; Chen, S.-M.; Piraman, S. The electrochemical synthesis of Pt particles on ZrO_2–ERGO modified electrodes with high electrocatalytic performance for methanol oxidation. *New J. Chem.* **2015**, *39*, 953–961. [CrossRef]
28. Ji, P.; Wang, Z.; Shang, X.; Zhang, Y.; Liu, Y.; Mao, Z.; Shi, X. Direct Observation of Enhanced Raman Scattering on Nano-Sized ZrO_2 Substrate: Charge-Transfer Contribution. *Front. Chem.* **2019**, *7*, 245. [CrossRef]
29. Jiao, X.; Chen, D.; Xiao, L. Effects of organic additives on hydrothermal zirconia nanocrystallites. *J. Cryst. Growth* **2003**, *258*, 158–162. [CrossRef]
30. Qin, D.; Chen, H. The influence of alcohol additives on the crystallization of ZrO_2 under hydrothermal conditions. *J. Mater. Sci.* **2006**, *41*, 7059–7063. [CrossRef]
31. Heuer, A.H.; Claussen, N.; Kriven, W.M.; Ruhle, M. Stability of Tetragonal ZrO_2 Particles in Ceramic Matrices. *J. Am. Ceram. Soc.* **1982**, *65*, 642–650. [CrossRef]
32. Yang, L.; Jiang, X.; Ruan, W.; Zhao, B.; Xu, W.; Lombardi, J.R. Observation of Enhanced Raman Scattering for Molecules Adsorbed on TiO_2 Nanoparticles: Charge-Transfer Contribution. *J. Phys. Chem. C* **2008**, *112*, 20095–20098. [CrossRef]
33. Toraya, H.; Yoshimura, M.; Somiya, S. Calibration Curve for Quantitative Analysis of the Monoclinic-Tetragonal ZrO_2 System by X-Ray Diffraction. *J. Am. Ceram. Soc.* **2010**, *67*, C-119–C-121.
34. Kim, B.K.; Hahn, J.W.; Han, K.R. Quantitative phase analysis in tetragonal-rich tetragonal/monoclinic two phase zirconia by Raman spectroscopy. *J. Mater. Sci. Lett.* **1997**, *16*, 669–671. [CrossRef]
35. Xu, J.P.; Wang, J.F.; Lin, Y.B.; Liu, X.C.; Lu, Z.L.; Lu, Z.H.; Lv, L.Y.; Zhang, F.M.; Du, Y.W. Effect of annealing ambient on the ferromagnetism of Mn-doped anatase TiO_2 films. *J. Phys. D Appl. Phys.* **2007**, *40*, 4757–4760. [CrossRef]
36. Zhao, X.; Vanderbilt, D. Phonons and lattice dielectric properties of zirconia. *Phys. Rev. B* **2002**, *65*, 075105. [CrossRef]
37. Carlone, C. Raman spectrum of zirconia-hafnia mixed crystals. *Phys. Rev. B Conden. Mat.* **1992**, *45*, 2079–2084. [CrossRef]
38. Duc Huy, L.; Laffez, P.; Daniel, P.; Jouanneaux, A.; The Khoi, N.; Siméone, D. Structure and phase component of ZrO_2 thin films studied by Raman spectroscopy and X-ray diffraction. *Mater. Sci. Eng. B* **2003**, *104*, 163–168. [CrossRef]
39. Yang, L.; Zhang, Y.; Ruan, W.; Zhao, B.; Xu, W.; Lombardic, J.R. Improved surface-enhanced Raman scattering properties of TiO_2 nanoparticles by Zn dopant. *J. Raman Spectrosc.* **2010**, *41*, 721–726.
40. Chang, S.-m.; Doong, R.-a. Interband Transitions in Sol-Gel-Derived ZrO_2 Films under Different Calcination Conditions. *Chem. Mater.* **2007**, *19*, 4804–4810. [CrossRef]
41. Myrick, M.L.; Simcock, M.N.; Baranowski, M.; Brooke, H.; Morgan, S.L.; McCutcheon, J.N. The Kubelka-Munk Diffuse Reflectance Formula Revisited. *Appl. Spectrosc. Rev.* **2011**, *46*, 140–165. [CrossRef]
42. Orendorff, C.J.; Gole, A.; Sau, T.K.; Murphy, C.J. Surface-Enhanced Raman Spectroscopy of Self-Assembled Monolayers: Sandwich Architecture and Nanoparticle Shape Dependence. *Anal. Chem.* **2005**, *77*, 3261–3266. [CrossRef]
43. Jiang, L.; You, T.; Yin, P.; Shang, Y.; Zhang, D.; Guo, L.; Yang, S. Surface-enhanced Raman scattering spectra of adsorbates on Cu_2O nanospheres: Charge-transfer and electromagnetic enhancement. *Nanoscale* **2013**, *5*, 2784–2789. [CrossRef]
44. Richter, A.P.; Lombardi, J.R.; Zhao, B. Size and Wavelength Dependence of the Charge-Transfer Contributions to Surface-Enhanced Raman Spectroscopy in Ag/PATP/ZnO Junctions. *J. Phys. Chem. C* **2010**, *114*, 1610–1614. [CrossRef]
45. Wang, X.; Li, P.; Han, X.X.; Kitahama, Y.; Zhao, B.; Ozaki, Y. Enhanced Degree of Charge Transfer in Dye-sensitized Solar Cells with a ZnO-TiO_2/N3/Ag Structure as Revealed by Surface-enhanced Raman Scattering. *Nanoscale* **2017**, *9*, 15303–15313. [CrossRef]
46. Chu, Q.; Li, J.; Jin, S.; Guo, S.; Park, E.; Wang, J.; Chen, L.; Jung, Y.M. Charge-Transfer Induced by the Oxygen Vacancy Defects in the Ag/MoO_3 Composite System. *Nanomaterials* **2021**, *11*, 1292. [CrossRef]
47. Yang, Y.; Li, J.; Zhang, M.; Song, P.; Lu, X.; Ding, Y. Universal and simple MoO_3 substrate for identification of SERS enhancement mechanism. *J. Raman Spectrosc.* **2021**, *1*, 1–6.
48. Ding, S.-Y.; Yi, J.; Li, J.-F.; Ren, B.; Wu, D.-Y.; Panneerselvam, R.; Tian, Z.-Q. Nanostructure-based plasmon-enhanced Raman spectroscopy for surface analysis of materials. *Nat. Rev. Mater.* **2016**, *1*, 16021. [CrossRef]
49. Xue, X.; Ji, W.; Mao, Z.; Mao, H.; Wang, Y.; Wang, X.; Ruan, W.; Zhao, B.; Lombardi, J.R. Raman Investigation of Nanosized TiO_2: Effect of Crystallite Size and Quantum Confinement. *J. Phys. Chem. C* **2012**, *116*, 8792–8797. [CrossRef]

50. Wang, Y.; Ji, W.; Sui, H.; Kitahama, Y.; Ruan, W.D.; Ozaki, Y.; Zhao, B. Exploring the Effect of Intermolecular H-Bonding: A Study on Charge-Transfer Contribution to Surface-Enhanced Raman Scattering of p-Mercaptobenzoic Acid. *J. Phys. Chem. C* **2014**, *118*, 10191–10197. [CrossRef]
51. Fu, X.; Jiang, T.; Zhao, Q.; Yin, H. Charge-transfer contributions in surface-enhanced Raman scattering from Ag, Ag_2S and Ag_2Se substrates. *J. Raman Spectrosc.* **2012**, *43*, 1191–1195. [CrossRef]
52. Guo, L.; Mao, Z.; Jin, S.; Zhu, L.; Zhao, J.; Zhao, B.; Jung, Y.M. A SERS Study of Charge Transfer Process in Au Nanorod-MBA@Cu_2O Assemblies: Effect of Length to Diameter Ratio of Au Nanorods. *Nanomaterials* **2021**, *11*, 867. [CrossRef]
53. Lin, J.; Hao, W.; Shang, Y.; Wang, X.; Qiu, D.; Ma, G.; Chen, C.; Li, S.; Guo, L. Direct Experimental Observation of Facet-Dependent SERS of Cu_2O Polyhedra. *Small* **2018**, *14*, 1703274. [CrossRef]
54. Lombardi, J.R.; Birke, R.L. Theory of Surface-Enhanced Raman Scattering in Semiconductors. *J. Phys. Chem. C* **2014**, *118*, 11120–11130. [CrossRef]
55. Wang, X.; Shi, W.; She, G.; Mu, L. Surface-Enhanced Raman Scattering (SERS) on transition metal and semiconductor nanostructures. *Phys. Chem. Chem. Phys.* **2012**, *14*, 5891–5901. [CrossRef] [PubMed]
56. Gritsenko, V.; Gritsenko, D.; Shaimeev, S.; Aliev, V.; Nasyrov, K.; Erenburg, S.; Tapilin, V.; Wong, H.; Poon, M.C.; Lee, J.H.; et al. Atomic and electronic structures of amorphous ZrO_2 and HfO_2 films. *Microelectron. Eng.* **2005**, *81*, 524–529. [CrossRef]

 *nanomaterials*

Article

DFT and TD-DFT Investigation of a Charge Transfer Surface Resonance Raman Model of N3 Dye Bound to a Small TiO_2 Nanoparticle

Ronald L. Birke * and John R. Lombardi

Department of Chemistry and Biochemistry, The City College of the City University of New York, 160 Convent Avenue, New York, NY 10031, USA; jlombardi@ccny.cuny.edu
* Correspondence: rbirke@ccny.cuny.edu

Citation: Birke, R.L.; Lombardi, J.R. DFT and TD-DFT Investigation of a Charge Transfer Surface Resonance Raman Model of N3 Dye Bound to a Small TiO_2 Nanoparticle. *Nanomaterials* **2021**, *11*, 1491. https://doi.org/10.3390/nano11061491

Academic Editor: Sergio Brutti

Received: 5 May 2021
Accepted: 28 May 2021
Published: 4 June 2021

Abstract: Raman spectroscopy is an important method for studying the configuration of Ru bipyridyl dyes on TiO_2. We studied the [Ru(II)(4,4′-COOH-2,2′-bpy)2(NCS)2)] dye (N3) adsorbed on a $(TiO_2)5$ nanoparticle using Density Functional Theory, DFT, to optimize the geometry of the complex and to simulate normal Raman scattering, NRS, for the isolated N3 and the N3–$(TiO_2)5$ complex. Two configurations of N3 are found on the surface both anchored with a carboxylate bridging bidentate linkage but one with the two NCS ligands directed away from the surface and one with one NSC tilted away and the other NCS interacting with the surface. Both configurations also had another –COOH group hydrogen bonded to a Ti-O dangling bond. These configurations can be distinguished from each other by Raman bands at 2104 and 2165 cm^{-1}. The former configuration has more intense Normal Raman Scattering, NRS, on TiO_2 surfaces and was studied with Time-Dependent Density Functional Theory, TD-DFT, frequency-dependent Raman simulations. Pre-resonance Raman spectra were simulated for a Metal to Ligand Charge Transfer, MLCT, excited state and for a long-distance CT transition from N3 directly to $(TiO_2)5$. Enhancement factors for the MLCT and long-distance CT processes are around 1 × 103 and 2 × 102, respectively. A Herzberg–Teller intensity borrowing mechanism is implicated in the latter and provides a possible mechanism for the photo-injection of electrons to titania surfaces.

Keywords: Raman; surface enhance Raman scattering; charge transfer; surface geometry; UV-VIS; DSSC; N3 and related Ru bipyridyl dyes

1. Introduction

The study of photoinduced processes at semiconductor-molecule interfaces has received much attention in the scientific literature because of its importance in both photocatalytic and solar energy conversion processes. Chief among these processes is photoinduced electron transfer at nanostructured titanium dioxide nanoparticle (NP) surfaces sensitized by adsorbed dye-molecules. This process is the basis of the dye-sensitized solar cell (DSSC), first characterized by Grätzel and O'Regan [1]. An important aspect of the properties of these solar cells is the nature of the adsorption geometry of the dye molecule on the surface of the NP TiO_2. The bonding of the adsorbed dye to Ti atoms at the titania surface affects the efficiency of the solar energy conversion process through both the injection time of electron transfer and changes in the conduction band structure at the surface of the TiO_2 [2,3]. Although many new sensitizer dyes have been studied to improve solar conversion efficiency [4], the prototype dye [2] for studying DSSCs has been cis-di(isothiocyanate)di(2,2′-bipyridine-4,4′-dicarboxylic acid)Ru(II) or $[Ru(NCS)_2(dcbpy)_2]$, known as N3 whose structure is shown in Figure 1. Neutral N3 in the solid form has all four carboxylates protonated but dissociates in water. A related dye is the N719 salt where two carboxylates in N3 have been deprotonated and tetrabutylammonium groups (TBA^+) give charge neutrality for the solid. In aqueous solution at pH $\geq$ 1.5 both compounds form the N3 dianion [5]. The Raman spectroscopy of N3 absorbed on a small TiO_2 nanoparticle by quantum mechanical electron structure methods is the subject of investigation in this paper. We will show that this model is able to give the most important binding geometry of N3 on TiO_2

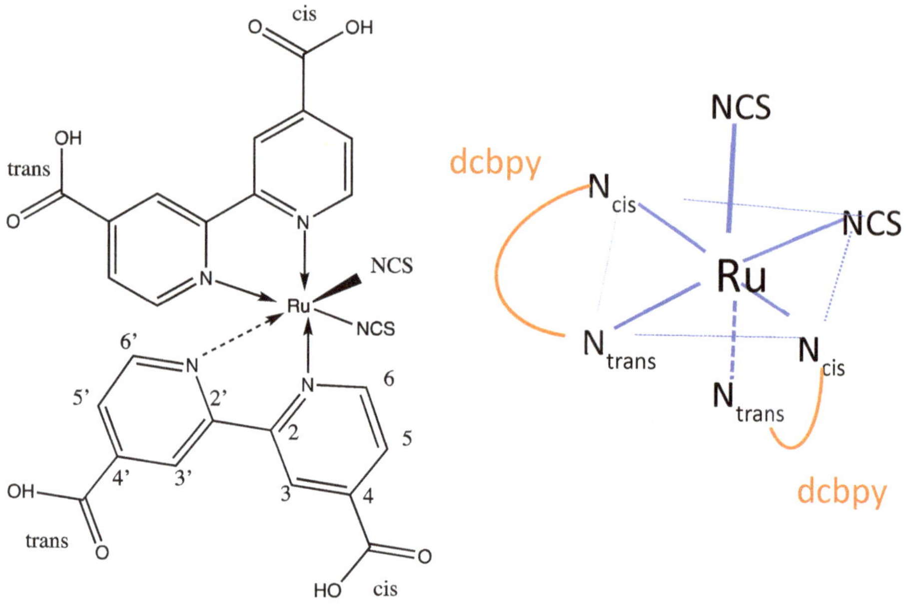

Figure 1. The chemical structure $Ru(NCS)_2(dcpby)_2$ called N3 and sketch of its structure.

Figure 1 displays a sketch of the pseudo-octahedral complex on the right showing the nitrogen atoms of the pyridyl rings and carboxyl groups of dicarboxylbipiridyl [dcbpy] which are cis and trans to the isothiocyanate ligands. There is a large number of studies of N3 and N719 dyes adsorbed on titania surfaces, and we review in some depth the pertinent literature related to Raman and Optical spectroscopy, and the dye-nanoparticle structure.

1.1. Raman and Infrared Studies

Both infrared and Raman vibrational spectroscopy methods are important experimental techniques for examining the structure of adsorbed molecules like N3 on TiO_2. Since normal Raman scattering is a relatively weak effect, some type of enhancement mechanism is necessary to obtain well-developed spectra for molecules on semiconductor surfaces. Indeed, adsorbed molecules on TiO_2 nanoparticles have been shown experimentally to support Surface Enhance Raman Scattering (SERS) via a charge transfer (CT) mechanism [6,7], i.e., CT-SERS. For the TiO_2-molecule system, it should be noted that a significant electromagnetic plasmonic field enhancement is unlikely as would be the case on metal substrates. For molecules on semiconductor (SC) surfaces, Raman enhancement mechanisms could involve either a solely molecular resonance Raman process for molecules which absorb light in the visible or a long-distance CT resonance mechanism for which the excitation occurs between the molecule and SC [8,9]. We will illustrate that these two photoinduced CT mechanisms are possible for N3 adsorbed on TiO_2 with molecular resonance from Franck-Condon scattering and long-distance CT between N3 and the TiO_2 nanoparticle from Herzberg-Teller scattering [8,9].

CT-SERS can be established as the enhancement mechanism [6] for adsorbed 4-mercaptobenzoic acid(4-MBA) on nanoparticle TiO_2 where a molecular resonance is not possible. Another possible photoinduced CT resonance mechanism for the TiO_2-molecule interface has been proposed [7] involving direct photoinduced charge transfer from the molecule ground state directly to the solid support without involving a Lowest Unoccupied Molecular Orbital, LUMO, on the molecule. This was called a Surface Enhanced Resonance Raman Scattering, SERRS, mechanism which should be distinguished from that of SERRS on metal substrates which involves molecular resonance in the molecule and the local surface plasmon resonance (LSPR) electromagnetic field enhancement from the metal substrate. Recently enhancement factors of 1.86×10^6 and 1.3×10^6 have been observed for 4-mercaptobenzoic acid on two-dimensional amorphous titania nanosheets [10] and for crystal violet on a nanofibrous three-dimensional TiO_2 network [11], respectively.

Of course, even higher enhancements can be obtained on SERS metal substrates like Ag and Au. A SERRS study of Ru dyes N719, N749, and Z907 was made on Au nanopeanuts and Au/Pt/Au nanorasberries [12]. This molecular resonance is due to Ru metal to ligand charge transfer, MLCT, which dominate the visible absorption spectra of the Ru bipyridyl dyes. An interesting feature of this study was the enhancement of the band at 2149 cm^{-1} from isothiocyanate indicating that N719 absorbs on Au via the sulfur end of the two NCS groups which are cis to each other. In an earlier study, Perez León obtained the spectra of N719 on Ag and Au colloidal nanoparticles in water, ethanol, and acetonitrile and assigned the bands in the different solvents [13]. N719 has also been studied on nanoporous Au surfaces [14]. The above studies show the difference between normal Raman scattering (NRS) spectra compared with resonance Raman scattering (RRS), SERS, and SERRS spectra on Ag and Au NPs.

Various types of surface Raman studies of N3 or N719 dyes have also been made on Ag film substrates which contain TiO_2 layers, such as Ag@TiO_2 core shell NP structures where the dye is on the titania surface [15,16]. Excitation at 532 nm includes both an intramolecular resonance and the LSPR enhancement [15]. Potential-dependent studies indicate that NCS is interacting with the TiO_2 surface through the S atom. Also, the direct photoinduced CT from the Highest Occupied Molecular Orbital, HOMO, of N719 to the TiO_2 conduction band is indicated [16]. Bing Zhao and coworkers [17] investigated Ag/N719/ TiO_2 sandwich systems excited at 532, 633, and 785 nm with N719 adsorbed to Ag NPs via the isothiocyanate groups. A similar study was made for Ag/N3/TiO_2 and TiO_2/N3 systems [18]. Here the 1366 cm^{-1} band indicates that N3 is connected to TiO_2 via the carboxylate group of the bipyridine groups. Two configurations of adsorbed N3 on TiO_2 particles are proposed. One solely bound by two trans carboxylates groups and another bound with one cis carboxylate group and one cis isothiocyanate group.

Vibrational spectroscopy without SERS metal enhancements on clean TiO_2 surfaces has also been used extensively to examine the N3 class of Ru sensitizer dyes with the goal of understanding how these dyes are coordinated to the surface of TiO_2 through the carboxylate group. The three types of possible surface coordination which have been discussed in the literature are shown in Figure 2. One of the first studies with Raman and Infrared, IR, spectroscopy of N3 and its deprotonated forms on nanocrystalline titania was made by Finnie et al. [19]. They ruled out the monodentate configuration and concluded that N3 attaches to the TiO_2 surface via either a bidentate chelate or bridging bidentate coordination (Figure 2) based on an empirical correlation in the literature [20] between the splitting of symmetric and asymmetric stretching vibrations for ionic carboxylate and N3 adsorbed on TiO_2. Also, each N3 molecule was considered to bind to the surface with two coordinating carboxyl groups [19]. Bands in solid N3, N719, and N712 have also been assigned with Fourier Transform Infrared, FTIR in the photoacoustic mode [21]. For these dyes on TiO_2 films, Attenuated Total Reflectance, ATR-FTIR, spectra were also obtained.

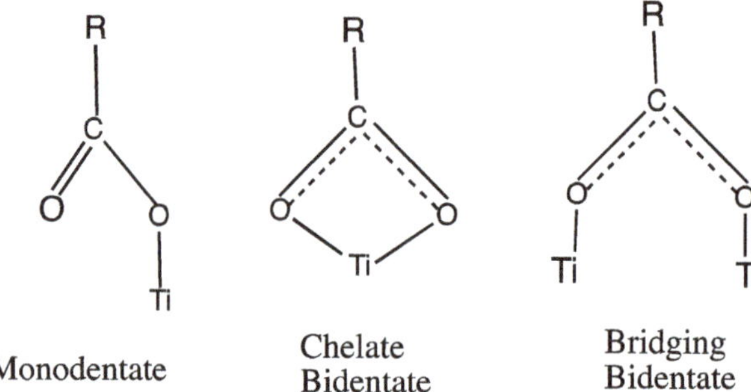

Figure 2. Three types of surface bonding to Ti atoms on titania surfaces.

In a later Raman study, well-defined spectra were obtained by Greijer et al. [22] from the anode of solar cells made by soaking nanostructured TiO_2 with N719 dye. In their work the emphasis was on the reaction of iodine with the isothiocyanate groups of the dye. The study of Shoute and Loppnow [23] obtained resonance Raman spectra at 514.5 nm excitation for free N3 in dimethyl sulfoxide, DMSO, solution and for N3 adsorbed on colloidal nanoparticles of TiO_2 dissolved in DMSO. The dye was considered to be covalently bound to the Ti(iV) surface ions by the carboxylate groups on the bipyridine ligand. The resonance process of a metal-to-ligand charge transfer process, MLCT, was considered to occur by photoexcitation of N3 with oxidation of Ru(II) to Ru(III) and the formation of a bipyridine radical anion with subsequent fast electron transfer via carboxylate π^* orbitals to the d orbitals of Ti of the conduction band. Shoute and Loppnow attribute the 1388 cm^{-1} band to a bridging or bidentate chelate linkage. The binding of Ru-bpy dye sensitizers on mesoporous TiO_2 was again investigated with Raman and FTIR spectroscopies by Perez Leòn et al. [24]. For the N719, they concluded again that the anchoring occurs by either bidentate chelate or bidentate bridging coordination. This adsorption geometry for N719 is in agreement with Finnie et al. [19] and Nazeeruddin et al [21]. A later study by Lee et al. [25] concluded that additional consideration of surface groups of titania such as Ti-O, Ti-OH, and Ti-OH_2 were necessary in considering the surface geometry of –COO^- and –COOH anchoring in dyes such as N719.

In the above studies, the use of Raman and IR results could not distinguish between the bidentate chelate and the bridging bidentate modes of surface anchoring which stimulated further vibrational spectroscopy studies of N719 on TiO_2 films. Schiffmann et al. [26] focused on the nature of the carboxylate binding geometry on TiO_2 using ATR-FTIR experiments and DFT theoretical simulations. The IR results were compared with electronic structure calculations made with a slab model and a periodic boundary hybrid Gaussian and plane-wave (GPW) DFT methodology using CP2K software. It was found that none of the eleven vacuum optimized dye-TiO_2 surface structures had the bidentate chelate mode. Our results, to be presented herein, also did not show this bidentate chelate geometry.

The consideration from the above vibrational and DFT calculational studies of N3 like Ru dyes indicate that several adsorption configurations are possible and that they may coexist or can be interconverted. The Raman and IR spectra seem to rule out the monodentate ester configuration, but could not differentiate between the bridging bidentate or bidentate chelate forms [21,24]. The most definitive vibrational marker of chemisorption via carboxylate bonds for N3-like sensitizer dye is the band in 1370–1388 cm^{-1} region assigned to the symmetric COO^- stretch since it is found for N3 on the TiO_2 film but not in solution. Raman simulations with our N3-$(TiO_2)_5$ model agree with this conclusion.

One of the first approaches to modelling the adsorption configuration of the N3 dye on TiO_2 was to consider the known crystal structure of nanocrystalline anatase (101) in relationship to the determined X-ray crystal structure of N3 [27]. Several adsorption models were proposed based on carboxylate interaction with the anatase (101) cut surface. This paper is an important source of data for comparing bond distance and angles from DFT calculations with X-ray data for N3.

1.2. Optical Studies, Theoretical Electronic Structure, and Charge Transfer Mechanisms

In order to elucidate Raman behavior of N3 and N719 dyes and indeed, the fundamental processes in dye-sensitized solar cells, it is important to model the UV-VIS absorption curves for the isolated dyes and the dyes adsorbed on model TiO_2 NP surfaces. Fantacci and co-workers [28,29] carried out DFT geometry optimization and TD-DFT absorption spectra simulations for isolated N3 in vacuum and in ethanol. Persson and Lundqvist [30] introduced a model for N3 adsorbed on a $(TiO_2)_{38}$ 1.5 nm nanoparticle cut from (101) anatase and carried out optical absorption simulations. Full geometric optimization at the B3LYP level in vacuum were calculated with Ru, N, and S atoms using the LANL2DZ ECP basis sets while 6–31G* was used for H, C, O, and Ti atoms. The structure of this complex had two binding sites to the nanocrystal from carboxylates on the same bipyridine and was titled so that S atom end of the isothiocyanate was 3.5 Å from the surface. The optimized structure had two bridge binding anchoring groups. These calculations show that the first few excitations involve MLCT transitions to orbitals within a few tenths of an eV of the conduction band orbitals.

This work was followed by several papers on electronic structure calculations from Grätzel and coworkers [3,31–33] with N719 dye also on $(TiO_2)_{38}$ nanocrystal model surfaces. In these papers, the structures of N719 and other related dyes adsorbed on TiO_2 slab models were optimized by the Car-Parrinello (CP) method. With two anchoring carboxylate groups, a structure called B with two protons transferred to the surface was energetically favored [3,31] over the structure A where the protons were retained on the dye. A most interesting results is that on the low energy structure B [31], there are mixed-dye TiO_2 states. The authors conclude this strong coupling suggests ultrafast electron injection rates. These studies illustrate that mixed states could be considered as indicating a direct electron transfer from the Ru-NCS HOMO in the TiO_2 band gap to empty conduction band orbitals of the TiO_2 cluster [31,33]. In fact, a charge density difference plot shows this direct electron transfer [33]. Such a direct excitation mechanism is also indicated from the natural transition orbital (NTO) hole-electron plot we will present herein.

For DSSCs there are a number of kinetic pathways which are important [34]. However, for the surface Raman enhancement process for Ru sensitizer dyes on the TiO_2 surfaces, it is the mechanism of the photo-induced electron transfer, ET, which is of primary interest. Such a mechanism could change with a change in the excitation energy of the exciting light. Two possible types of ET mechanisms for the dye-TiO_2 interface have been delineated in the literature [35–37]. Person et al. [35] described the ET mechanism on the basis of INDO semiempirical calculations for a dye-TiO_2 model. In both types of ET, the ground state of the dye lies in the gap between the valence band (VB) edge and the conduction band (CB) edge. In scheme 1, the photoexcitation transfers the electron from the dye HOMO to an antibonding unoccupied intermediate state above the conduction band. It is from this level that the ET injection occurs to the Ti(3d) levels of the conduction band, presumably via electron tunneling. In scheme 2, the photoexcitation transfers the electron from the dye HOMO directly to the conduction band edge, without going through an intermediate excited state of the dye. This latter scheme is facilitated by a strong substrate-adsorbate interaction. For surface Raman scattering at TiO_2, both schemes 1 and 2 are possible and would engender a different enhancement process: a molecular resonance Raman process (Franck-Condon scattering) in scheme 1 and a long distance CT from dye-to-TiO_2 (Herzberg-Teller scattering) in scheme 2, where intensity borrowing is possible [8,9]. For DSSCs these electron injection schemes were called Type-I (two-step) and Type-II (one-step) pathways, and the Ru(II) dyes bound to TiO_2 through carboxylates were classified as Type-1 [36]. This classification is certainly dependent on whether the energy of the exciting light promotes the electron to a state on the dye or to the Ti(3d) unoccupied state of titania. Thus, for excited state S18 in the model of De Angelis et al. [33], N719-$(TiO_2)_{38}$ follows a scheme 2

(Type-II) one-step electron injection. However, early experimental results were considered to show the two-step mechanism (Type -I) [38]. The injected electron rate is considered to be controlled by the Franck-Condon overlap of vibrational states of reactant and product. Another study of dyes on colloidal TiO_2 [39] using Stark emission spectroscopy found evidence for both types of mechanisms indicating the possibility of injection from a state with significant interfacial CT character. Our excited state hole-electron iso-surface plot also suggests such a one-step direct CT process is possible.

The two-step mechanism has been investigated in a theoretical study of resonance Raman scattering, RRS, at a dye-semiconductor surface [40]. This treatment is based on an Anderson-Newns type Hamiltonian with a continuum of semiconductor states approximated by a set of discrete states with uniform energy in terms of Legendre polynomials. It was found that for this model intensities of Raman active modes in the isolated dye are de-enhanced on the surface because of the population decay of the dye excited state by the electron injection process. A one-step mechanism would not be subjected to this decay, but here one would expect to observe a red shifted optical spectrum from a CT state with a dye cation D^+ and the electron e^- in the conduction band. However, if the oscillator strengths for excitation to such CT states are very weak, they would not be observable in the absorption spectrum. In fact, both the direct one-step and intermediate two-step electron transfer mechanisms for different configuration of xanthene molecule complexes on the same TiO_2 surface has been proposed [41].

In several ultrafast experiments, typically at 400 nm excitation [42], the two-step process has been assumed with excitation to the MLCT state of N3 with the electron in the π^* orbital of dcbpy followed by injection of the electron to the semiconductor conduction band. These authors conclude that a direct CT transition is unlikely from Ru dye to TiO_2 because of the lack of overlap between Ru dye and TiO_2 orbitals. However, our calculations show some oscillator strength for such a transition.

Other evidence for a direct charge-transfer mechanism from adsorbed Ru dyes comes from Raman spectra with enhanced phonon modes of TiO_2 NPs [43–45]. With N719 on the TiO_2, there is enhancement of the phonon modes with 514.5 nm excitation which is attributed to the charge transfer process [44]. Fourth order coherent Raman spectroscopy with 20 fs pulses was also used to examine N3 dye on TiO_2 (110) surfaces [45]. This technique examines the interfacial region and thus surface modes of the TiO_2 can be observed. The surface mode at 100 cm^{-1} and the $E_g(1)$ mode at 146 cm^{-1} where found to be strongly enhanced. Several enhancement mechanisms were proposed including the Herzberg-Teller CT-SERS mechanism of Lombardi and Birke [8,9] where the ground state of N3 is located in between the valance band and conduction band and there is direct CT from this state to the conduction band with intensity borrowing from a strongly allowed transition of the dye.

1.3. *TiO_2 Nanoparticle Models*

Many studies in the literature have been made with models N3 or N719 molecule adsorbed on a TiO_2 NP complex using a $(TiO_2)_{38}$ model for the titania nanoparticles [3,31–33,35]. Even larger TiO_2 model NPs have also been explored [46,47]. An important parameter for assessing the adequacy of the $(TiO_2)_n$ clusters is the HOMO-LUMO (HL) gap, which is an estimate of the band gap between the valence and conductance band edges. With B3LYP/LANL2DZ, for n between 16 and 60, band gaps between 4.55 and 4.94 eV were determined [48]. In another study [47], B3LYP and a variety of basis sets were used to calculate the HL gap for n = 1 to 68. Because we are calculating static and time-dependent Raman with TD-DFT, a large nanocluster model was found impractical for our calculations. We have thus used a much smaller $(TiO_2)_n$ cluster with n = 5. For this nanocluster, our B3LYP/6-31+G(d)/LANL2DZ result for the HOMO-LUMO (HL) gap is 4.45 eV which is within 0.1 eV of calculated values with a B3LYP/6-311 + G(2df,p) calculation for the $(TiO_2)_5$ cluster C in the paper of Lundqvist et al. [47]. Furthermore, HL band gap for n = 5 found herein is consistent with the HL band gap for much larger nanoclusters up to n = 68 [47].

This HL band gap is of the order of 1.5 eV larger than the experimental band gap or a bulk band gap calculated using a B3LYP and a periodic calculation. This increased band gap found for nanoclusters up to about 1nm in size has been attributed to a quantum size effect (confinement). Furthermore, there was a good quantitative fit of excitations energies for TD-B3LYP compared with the Equation of Motion Coupled-Cluster, EOM-CC, method [48]. Thus, based on the reasonable results for B3LYP properties of the $(TiO_2)_5$ NP for ground state and excited state calculations, we have used this size, n = 5, cluster with N3 adsorbed on the surface to simulate our Raman calculations.

2. Computational Methods and Models

Various forms of N3 dye, cis-Ru(II)-$(dcbpy)_2(NCS)_2$, either isolated or as a complex bound to a $(TiO_2)_5$ nanoparticle were used for spectral simulations. Calculations were made with DFT and TD-DFT using the Gaussian 16 software package [49] for geometric optimization, IR and Raman frequency, and time-dependent excited state optical and Raman calculations. All ground and excited state frequency calculations were made in the harmonic approximation. All calculations were made with the B3LYP hybrid GGA exchange-correlation density functional [50,51] with the 6-31+G(d) basis set for C, N, O, S, and H and with the LANL2DZ effective core potential (ECP) basis set for the Ru and Ti atoms. The B3LYP density functional has been used in most studies of dyes on TiO_2 surfaces. UV-VIS absorption spectra were calculated for both gas phase (vacuum) and an ethanol solvent environment. For the ethanol calculations, the optimized structure and excitation spectrum were made with the SCRF default method of G16, which is the Polarizable Continuum Model (PCM) using the integral equation formalism variant (IEFPCM). The nature of the excited state transitions for Gaussian calculations was examined with Natural Transition Orbitals (NTO) [52] and with the charge-transfer distance index, D_{CT}, and the amount of charge transferred index q_{CT} [53].

Normal Raman vibrational frequencies with G16 were computed at a stationary point by determining the second derivatives of the energy with respect to the Cartesian nuclear coordinates and then transforming to mass-weighted coordinates. In all cases, no imaginary frequencies were found. Pre-resonance frequency-dependent TD-DFT Raman vibrational calculations were made in G16 with the Coupled Perturbed Hartree Fock (CPHF) method. An empirical scaling factor of 0.970 was used to scale all normal mode wavenumbers.

To simulate Raman spectra with an N3–TiO_2 nanoparticle complex, we have used N3 bound to a Ti_5O_{10} nanoparticle. This model is based on the reaction of a single negatively charged N3, which is formed by the loss of a proton from one of the four COOH groups on neutral N3, forming a bridging bidentate bond to the neutral Ti_5O_{10} nanoparticle. This single –COO carboxylate anchoring group of N3 forms a bridging bond to two Ti atoms of the Ti_5O_{10} nanoparticle. The structure of Ti_5O_{10} was cut from a much larger slab of an anatase polymorph. The initial structures of the N3–Ti_5O_{10} complex with unidentate ester or bidentate chelate binding to a single Ti atom were found to always converge to the bridging bidentate structure on this Ti_5O_{10} nanoparticle. Previous theoretical studies of dyes adsorbed on TiO_2 polymorphs have used much larger nanoparticle models but typically with a much lower level basis set such as 3-21G(d) and did not include Raman simulations. These were mostly made for the study of excited states. However, for accurate vibrational Raman calculations, a higher order basis set is necessary. Geometric optimization followed by Raman vibrational simulation for N3 bound to larger TiO_2 nanoparticle models with a B3LYP/6-31+G(d)/LANL2DZ calculation was found to be impractical. GaussView 6.0 [54], Chemcraft [55], and Multiwfn 3.8 [56] were used to view and analyze structures and spectra.

3. Results and Discussion

3.1. Optimized Geometry of Fully Protonated N3

The optimized geometry of all forms of N3 was found to be a pseudo-octahedral complex of Ru^{2+} as shown for the neutral species in vacuum in Figure 3. The Ru central atom is coordinated to six nitrogen atoms from two bipyridyl groups and two isothiocyanate ligands. Here, N3 optimizes to C_1 symmetry; however, in ethanol, the structure optimized to C_2 symmetry with B3LYP/6-31+G(d)/LANL2DZ. In these structures, as seen in Figure 1, the two thiocyanate anions, NCS^-, are cis to each, and the remaining nitrogen atoms of the bipyridines binding to Ru(II) are either cis or trans to the thiocyanate ligands. One can also denote the carboxylate groups on the pyridine rings of bipyridine as either cis or trans to the isothiocyanates.

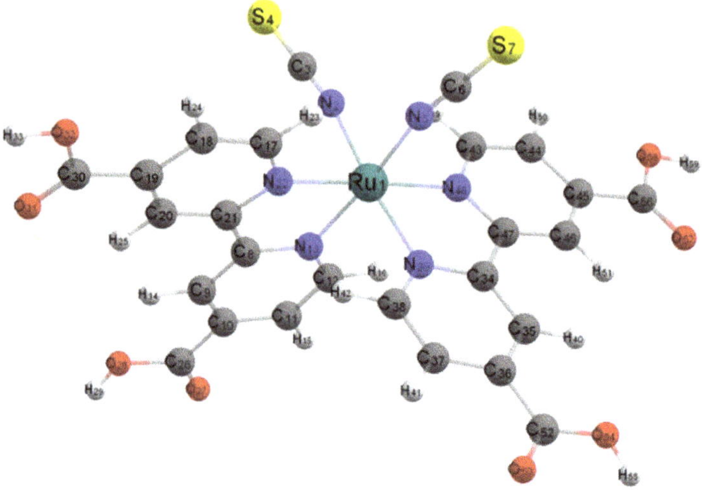

Figure 3. Optimized structure of neutral cis-N3 in vacuum.

In order to validate the computational method, we compare our calculations for neutral N3 complexes to two other vacuum calculations in the literature. Since there can be differences in the bond distance between the same type of Ru–N bonds on the different bipyridyl rings, we have denoted them as rings A and B. All the computational calculations in Table 1 show hardly any difference between rings A and B for the same type of Ru–N bond. However, this is not true for the X-ray experimental values. Our calculations are similar to the recent calculation [57] that was made with B3LYP and a slightly larger basis set, 6-311G(d) with a small core relativistic ECP for Ru. The three optimized structures in vacuum for N3 in Table 1 show slightly larger bond distances than the X-ray data [27] for Ru-bpy bonds; however, solid-state packing forces could account for these differences. By and large, all of the calculations in the table can be considered a reasonable representation of the N3 structure. Interestingly, the calculation with the BPW91 density functional [28] was constrained to C_2 symmetry, whereas our calculation in ethanol was optimized to C_2 symmetry.

The data above is for the fully protonated N3. We have also preformed geometric optimization in vacuum for the N3 monoanion, which is deprotonated from one cis-carboxylate, $N3^-$(cis), and for N3 monoanion, which is deprotonated from one trans-carboxylate, $N3^-$(trans), and for the dianion $N3^{2-}$ in which two cis-carboxylates are deprotonated, which are on different bipyridine rings. It should be pointed out that there is an approximate C_2 axis that bifurcates the angle formed by the two isothiocyanate groups and that the molecule can be considered as having one long axis and two short axes. The long axis is from the end of one cis-carboxylate to the end of the other cis-carboxylate [27]. The short axes are from the end of one isothiocyanate groups to the end of the opposite trans-carboxylate, as shown in Figure 3. These calculations for the anion structures were used to examine the effect of molecular charge on the dipole moment. For N3(0), $N3^-$(cis), $N3^-$(trans), and $N3^{2-}$(cis), the dipole moments are 12.62 D, 25.80 D,

15.56 D, and 59.68 D, respectively. For N3(0), the dipole moment is directed approximately along the C_2 axis with the negative end between the NCS^- groups. For $N3^-$(cis), the dipole moment is directed approximately along the long axis with the negative end on the deprotonated carboxylate. For $N3^-$(trans), the much lower dipole moment makes an approximately 90° angle with the C_2 axis and is approximately in the plane formed by the two isothiocyanates with its negative end near the Ru and its positive end directed over the other trans-carboxylate, which is protonated. For $N3^{2-}$, which is the same as the dianion of N719, the carboxylate groups are deprotonated on opposite sides of the long axis. Here, the dipole moment is also directed along the C_2 axis. It is interesting to compare this dipole moment, 59.68 D, with that of the similarly deprotonated species N719 b [3], 28.0 D, where the dipole moment is also along the C_2 axis. The difference is that in N719 b, the cis-carboxylate ions have Na^+ counterions, which balance the charge and lower the magnitude of the dipole moment by more than 50%. With N3(0) and $N3^-$(cis) in ethanol using the IEFPCM, self-consistent reaction field calculation, the dipole moments are 21.12 D and 38.50 D, showing the significant effect of solvation on increasing the electronic polarization. By and large, the direction and magnitude of the dipole moments follow chemical intuition.

Table 1. Bond distances in Å and bond angles for our calculated N3 structures compared with two others in the literature and data from an X-ray crystal structure of N3.

Parameters	Calc. B3LYP Vac.	Calc. B3LYP EtOH	Literature [58] B3LYP Vac.	Literature [28] BPW91 Vac. C_2	Experiment X-ray [27]
Ru-NCS					
Ru-NCS	2.058	2.086	2.045	2.036	2.048
Ru-NCS	2.058	2.086	2.045	2.036	2.046
Ru-N(bpy_{trans})A	2.085	2.086	2.077	2.079	2.036
Ru-N(bpy_{trans})B	2.085	2.086	2.076	2.079	2.058
Ru-N(bpy_{cis})A	2.080	2.094	2.074	2.056	2.030
Ru-N(bpy_{cis})B	2.081	2.094	2.074	2.056	2.013
N = C(NCS)	1.185	1.179	1.178		1.162–1.103
C-S (NCS)	1.628	1.647	1.626		1.615–1.685
<SCN-Ru-NSC	92.6	90.5		90.2	88.7(5)
<N(bpy_{trans})-Ru- N(bpy_{cis})	78.5	78.4		78.9	79.5(5)
<N(bpy_{cis})-Ru- N(bpy_{cis})	177.9	175.2		169.5	174.5(5)
<SCN-Ru-N(bpy_{rans})	172.7	173.8			
<N(bpy_{trans})-Ru-N(bpy_{trans})	93.9	92.6		95.1	90.6(5) 90.6(5)

3.2. Geometric Optimization of the N3-Ti_5O_{10} Complexes

The optimized structure of two N3–Ti_5O_{10} complexes is shown in Figure 4. Complex A was obtained by starting with the bridging bidentate structure, while complex B was obtained by starting with N3 bound by a monodentate ester linkage to one Ti atom. Both were optimized to the bridging bidentate structure through a trans-carboxylate group and both had a cis-carboxylate hydrogen-bond to a dangling O atom of the nanocluster; however, in complex B, one isothiocyanate is tilted toward the surface. This type of bridging anchoring was discussed as a possible arrangement for sensitizing dye on the (101) anatase surface where it was referred to as the B,C-type; see Figure 5 of reference [27]. The B-type is when a trans-carboxylate, which is on the so-called short molecular axis, binds to the surface, and the C-type is when cis-carboxylate, which is on the so-called long molecular axis, binds to the surface. Here, both complexes A and B in Figure 4 are anchored to the surface by the trans-carboxylate or B-type. For such a model anchoring geometry, 71 fragments in the Cambridge Structural Database gave average parameters of bridging Ti-O(dye), 2.03 Å; O–C, 1.26 Å; Ti–O–C, 135°; and O–C–O, 124°. The bond distances and bond angles for our two complexes have values remarkably close to these average values.

Thus, for complex A, we calculated values of bridging Ti-O(dye), 2.02 Å and 2.07 Å; O–C, 1.27 Å; Ti–O–C, 139°, O–C–O, 125°. For complex B, we calculated values of bridging Ti–O(dye), 2.04 Å and 2.09 Å; O–C, 1.27 Å; Ti–O–C, 129°, O–C–O, 126°. Our data indicate that one side of the bridging bidentate linkage is shorter than the other side. In the eleven structures of Schiffmann et al. [26], which are calculated for a slab model with 144 TiO_2

units, the typical bond lengths are 2.08 Å for the bridging bidentate linkage and 2.02 Å for the monodentate bonds. It can be seen in Figure 4 that in both A and B complexes, the cis-carboxylate on the same bipyridine that binds to the surfaces through its trans-carboxylate forms a hydrogen bond to a terminal O atom of the nanocluster. The H-bond distance, C–O–H–O, is 1.76 Å for complex A and 1.47 Å for complex B. In B, the bond distance between the S atom of –NCS and Ti is 2.52 Å. The data for our optimized structure for complex A and B show good agreement between experimental averages and the simulated bridging bidentate bond distances.

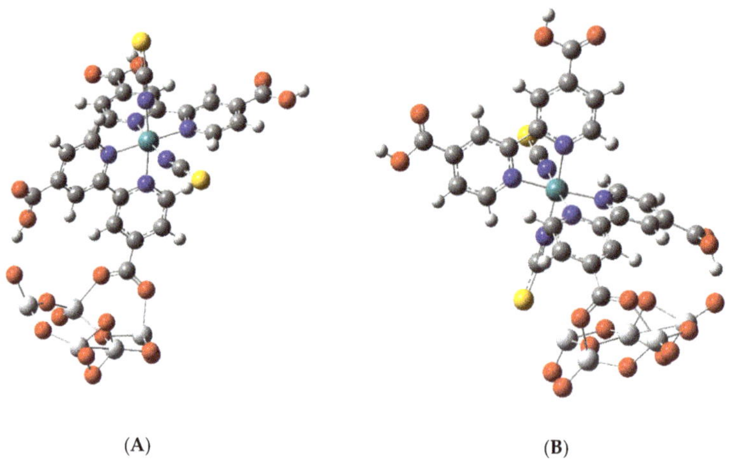

Figure 4. Optimized structures of two N3-$Ti5O_{10}{}^{1-}$ negatively charged singlet complexes. (**A**) Bridging bidentate structure with H-bond from a cis-carboxylate. (**B**) Bridging bidentate structure with a -NCS group bent toward the surface with H-bond from a cis-carboxylate.

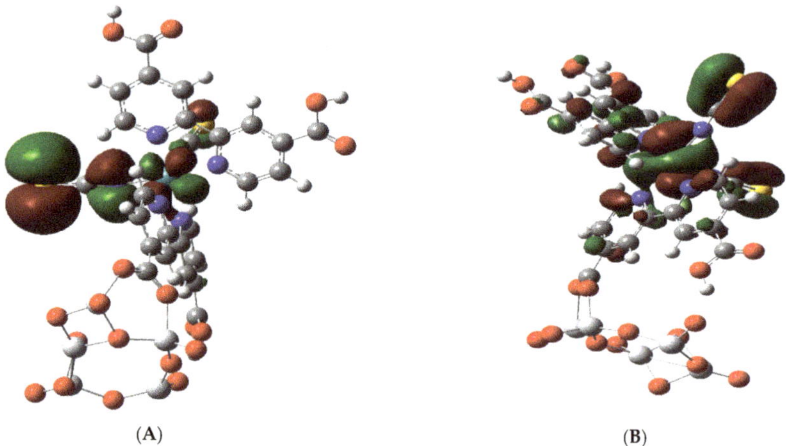

Figure 5. Filled molecular orbitals of the N3-$(TiO_2)_5{}^{-}$. Right Side (**A**): HOMO-1 which is similar to the HOMO and to H-2 and H-3 MOs. Left Side (**B**): HOMO-7 which is similar to H-5 and H-6 MOs.

As mentioned in the introduction our HOMO-LUMO (HL) gap for the optimized $(TiO_2)_5$ is 4.45 eV, which is within a few tenths of an eV calculated for much larger nanoclusters. Our optimized nanocluster has C_1 symmetry and contains three terminal Ti-O bonds with a 1.62 Å bond length and seven bridging O atoms. All Ti atoms in the structure are 4-fold coordinated, which is the same as a $(TiO_2)_5$ nanocluster of C_1 symmetry [58], which is 0.53 eV less stable than the global minimum. For comparison, the structure with the global minimum has an HL gap of 4.54 eV. When the carboxylate of N3 binds to the surface, it breaks a bridging O atom of the cluster, which creates one additional Ti-O terminal group in complexes A and B. We have calculated the binding energy for a singlet N3 anion reacting with neutral $(TiO_2)_5$ to give a minus one charged singlet species for the two complexes.

The binding energy E (binding) for an N3 monoanion reacting with the $(TiO_2)_5$ NP for complex A or B using the B3LYP functional is given by:

$$E[\text{binding}] = E[(TiO_2)_5] + E[N3(k)^-] - E[\text{complex}(i)^-] \quad (1)$$

where k = cis or trans deprotonated carboxylate and i = A or B complex, and all energies are optimized values in vacuum with a stationary point found after Raman analysis, which shows zero imaginary frequencies. The same basis set and G16 default grid size were used for all calculations, and values are not corrected for the basis set superposition error. For complex A, the binding energy was calculated for cis and trans species to be 72.23 kcal/mol and 70.47 kcal/mol, whereas for complex B, it was 93.16 kcal/mol and 91.54 kcal/mol, respectively. Thus, the binding energies are about the same for either cis or trans deprotonated $N3^-$ species. On the other hand, complex B is about 21 kcal/mol more stable than complex A. However, the normal Raman scattering (NRS) spectrum for complex A is calculated to be twice as intense as that for complex B, so that it would dominate Raman scattering if both occurred. Thus, for correlating the Raman spectrum with structure on titania surfaces, complex A is considered the more important structure.

3.3. Electronic Structure of the Complex

We will mostly be concerned with complex A, which we will refer to as $N3\text{–}(TiO_2)_5^-$. This complex has an electronic structure with a set of 13 molecular orbitals that span an energy region, which can be considered to correspond to the HL band gap of the Ti_5O_{10} NP. These molecular orbitals are listed in Table 2. We note that this complex has been constructed as a singlet anion assuming that the N3 anion with a deprotonated carboxylate binds with the neutral Ti_5O_{10}. Thus, it is interesting that the molecular orbitals of the N3 part of the complex closely resemble those of neutral N3 and not those of the N3 anion. However, this fact is consistent with the high positive charge on the two Ti atoms, which bind the two negatively charged O atoms on the carboxylate anchor of the dye and act similar to the protons in the free N3. Thus, the HOMO and the HOMO-1 to HOMO-7 orbitals match closely the same eight orbitals of free N3 dye [28], which is a fact that has been previously observed for the combined system [30]. The HOMO and the seven MOs below it contain mixed Ru 4d and isothiocyanate orbitals [28,30,59]; see Table 2. Since the Ru-N bond distances are unequal and the octahedral structure is distorted in the complex, similar to the free N3 (Table 1), the Ru 4d orbitals do not show the typical ligand field splitting of a d^6 ion in an octahedral field.

In fact, the calculated $N(bpy_{trans})$-Ru- $N(bpy_{cis})$ angle in the complex is 78.9°, which is similar to 78.5° calculated for N3 and to 79.5° for the X-ray structure of N3 (Table 1), showing a strong distortion from 90° of the octahedral structure. In addition, the SCN-Ru-$N(bpy_{trans})$ angle, 173°, in the complex deviates from 180° as it does in N3. These two types of L-M-L distortions of the octahedral geometry split the degeneracy of the t_{2g} set [60], as seen in Table 2.

In fact, not one of the six ligand-donating orbitals is along the z-axis, and they are also located in between the axes. A Natural Bond Orbital, NBO, analysis of N3 shows the natural atomic orbital energies and occupancies of the d orbitals as $4d_{x2\text{-}y2}$, −6.34 eV (1.73); $4d_{xz}$, −6.07 eV (1.50); $4d_{z2}$, −5.95 eV (1.53); $4d_{yz}$, −5.54 eV (1.15); $4d_{xy}$, −5.49 eV, (1.18). The natural bond orbital analysis shows two strong NBO bonds for the two Ru-NCS bonds with 1.97 occupation and 75% of the electron density on the sp N and 25% on the Ru hybrid. The three lone pair Ru occupations are: LP(1) (1.91), LP(2) (1.84), and LP(3) (1.73).

This NBO analysis shows the splitting of the d atomic orbital block and the polar covalent character of the six dative bonds between the ligands and Ru in this 18-electron distorted octahedral system. We have also examined bond orders for these Ru-N bonds. The Wiberg bond indices from the NBO analysis for isolated N3 are 0.625 for two Ru-N (NCS) bonds, 0.544 for two Ru-N (bpy_{cis}) bonds, and 0.554 for two Ru-N(bpy_{trans}) bonds. In addition, we calculated Mayer bond orders for Ru-N bonds in the N3-$(TiO_2)_5^-$ complex with Multiwfn software [57]. These are 0.754 for the Ru-N (NCS) bond trans to the anchoring carboxylate group, 0.747 for the other Ru-N (NCS) bond, 0.666 for the Ru-N (bpy_{cis}) and 0.672 for Ru-N(bpy_{trans}) bonds on the bpy with the carboxylate anchoring group, and 0.722 for Ru-N (bpy_{cis}) and 0.733 for Ru-N(bpy_{trans}) bonds on the other bpy ligand.

Table 2. Molecular orbitals in the putative band gap of the complex N3-$(TiO_2)_5^-$.

MO	Energy (eV)	Character
LUMO + 3	−1.34	Ti $3d_{Z^2}$ and π* dcbpy on both dcbpy
LUMO + 2	−1.67	π* dcbpy not bound to TiO_2 and Ru $4d_{xz}$
LUMO + 1	−1.92	π* dcbpy bound to TiO_2 and Ru $4d_{Z^2}$
LUMO	−2.19	π* dcbpy not bound, Ru $4d_{Z^2}$, and on bind. π* pyr-COO
HOMO	−3.72	S $3p_x$, N $2p_x$ on both NCS, Ru $4d_{yz}$
HOMO-1	−3.86	S $3p_x$, N $2p_x$ on both NCS, Ru $4d_{yz}$
HOMO-2	−3.94	S $3p_y$, N $2p_y$ on both NCS, Ru $4d_{yz}$
HOMO-3	−4.02	S $3p_y$, N $2p_y$ on one NCS, Ru $4d_{yz}$
HOMO-4	−5.36	S $3p_z$, N $2p_z$
HOMO-5	−5.41	C = S π on both NCS, Ru $4d_{Z^2}$
HOMO-6	−5.49	C = S π on both NCS, Ru $4d_{Z^2}$
HOMO-7	−5.57	C = S π on both NCS, Ru $4d_{Z^2}$
HOMO-8	−5.73	O 2p, Ti $3d_{xy}$

We find in the complex that below the HOMO, there is one set of three quasi-degenerate MOs HOMO-1 to HOMO-3 (H-1 to H-3). These can be represented by the iso-surface of HOMO-1 (Figure 5A), which contains a Ru $4d_{yz}$ orbital and p orbitals on the S and N atoms of NCS, which are anti-bonding with respect to each other. Below this set of three MOs, there is another set of four quasi-degenerate MOs from HOMO-4 to HOMO-7 (H-4 to H-7); see Table 2. HOMO-4 is unique for both the free N3 and the complex, since it does not contain a Ru orbital but only contains p orbitals on N and S atoms, which are also antibonding with respect to each other [28]. The three very similar molecular orbitals H-5 to H-7 contain a Ru 4 d_{Z^2} and a π bonding C = S combination and are represented by HOMO-7 (Figure 5B). Here, there is also some leakage around the Ru 4 d_{Z^2} orbital, which is bonding with the N atom of the NCS group. The HOMO-1, which is shown in Figure 5A, is also very similar to the HOMO and the HOMO-2 and HOMO-3 molecular orbitals.

The atomic orbitals for the TiO_2 NP first appear in molecular orbital HOMO-8 corresponding to the valence band edge. Thus, the HOMO-8 MO contains only 2p atomic orbitals of the O atoms and $3d_{xy}$ orbitals of Ti and is a pure MO of the TiO_2 nanocluster; see Figure 6A. The next four MOs below H-8 going down to H-12 at −6.28 eV are similar to it and thus are all pure TiO_2 MOs. Starting at HOMO-13, there is a mixed MO iso-surface which contains C = C π bonding in the bipyridine and O p orbitals on the cis COOH group of the bipyridine that binds to the NP as well as 2p orbitals on the O atom of the NP.

The LUMO + 3, which represents the conduction band edge, is a mixed MO with anti-bonding π* orbitals in both bipyridine rings and a Ti $3d_{Z^2}$ orbital; see Figure 6B. The energy difference between HOMO-8 and LUMO + 3 represents the approximate band gap and spans an energy of 4.39 eV. The total density of states (TDOS) and the partial density of states (PDOS) for Ru and Ti illustrating the putative band gap are shown in Figure 7. The approximate band gap can be seen in the Ti partial density of states plot (blue line). The putative band gap of the complex, 4.39 eV, is close to the 4.45 eV HL gap for the isolated $(TiO_2)_5$ NP. The HOMO at 3.72 eV is 2.02 eV above the putative valance band edge (H-8) and 2.37 eV below the putative conduction band edge (L + 3); see Figure 7. The HOMO, dotted line in Figure 7, is seen to be almost in the middle of the presumed band gap. There are three molecular orbitals above the HOMO and below LUMO + 3 in the putative band gap, which are the LUMO, LUMO + 1, and LUMO + 2, all containing anti-bonding π * C = C combinations of the bipyridine rings and unoccupied Ru 4d orbitals; see Table 2 and Figure 7. The LUMO + 2 is 0.3 eV below the "conduction band edge" MO (L + 3). These MO energy levels are similar to the calculations of Persson and Lundqvist [30], where the penultimate MO in the band gap was 0.2 eV below the conduction band edge in the much larger $(TiO_2)_{38}$ NP. Thus, the ground state electronic structure in our model is similar to the N3-$(TiO_2)_{38}$ model [30] except for the fact that our MO energies are shifted up ca. 1.5 eV because of the negative charge on the complex.

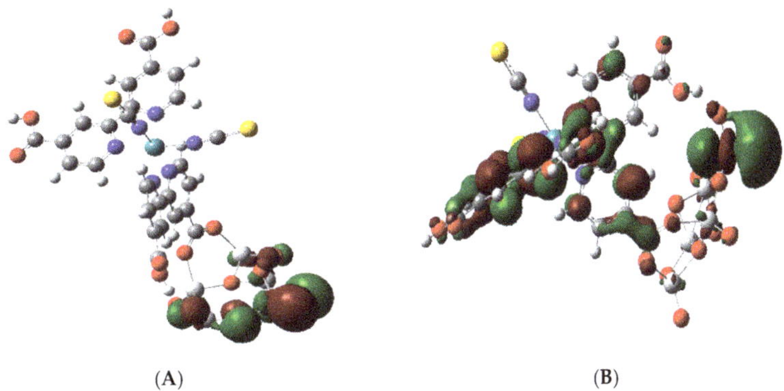

Figure 6. Molecular orbitals representing the band edges. (**A**): HOMO-8 showing atomic p orbitals of O atoms in $(TiO_2)_5$. (**B**): LUMO + 3 showing the $3d_Z{}^2$ orbital of Ti atoms of $(TiO_2)_5$.

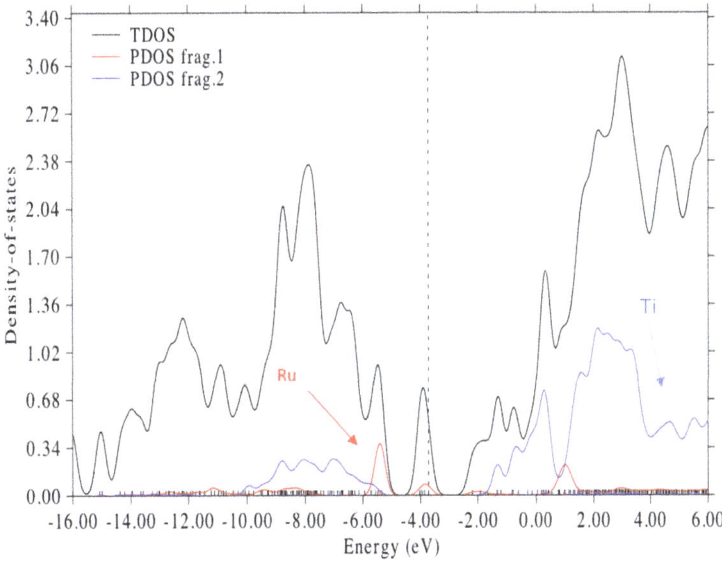

Figure 7. Total density of states (TDOS) and partial density of states (PDOS) for the N3–$(TiO_2)_5{}^{1-}$ complex. Frag. 1 (red) indicates states with the Ru atom. Frag 2 (blue) indicates states with Ti atoms. Gaussian broadening with 0.4 eV FWHM. Dotted line is at the HOMO energy 3.717 eV. Hirshfeld analysis was used for calculating orbital compositions with Multwfn 3.8 software [56].

In the N3-sensitized DSSCs, fast electron injection has been thought to involve MLCT transitions in a two-step process where the excited states of N3 are well above the conduction band of TiO_2 films [42]. However, it is the excitation energies and not the ground state unoccupied energy levels that are important for the photoinduced CT process. We discuss these excited states in the next section. Our N3–$(TiO_2)_5^-$ model complex appears to be similar to the electronic structure of the N3–$(TiO_2)_{38}$ model, and thus, it should be adequate to investigate optical absorption and Raman scattering spectra for a surface site with the bridging bidentate structure.

3.4. Simulated Optical Absorption Spectra

In Figure 8, we have plotted the TD-DFT simulated optical absorption curves for N3 and the complex A, N3-$(TiO_2)_5^-$, which are both obtained in vacuum. The two spectra are for the most part very similar. One difference in the two curves is that the shoulder at 250 nm for N3 is not reproduced for the complex. This is an artefact stemming from that the fact that even with 150 excited states, the B3LYP/6-31 + G(d)/LANL2DZ simulated spectrum does not quite go down to 240 nm where the N3 spectrum starts. This calculation involved 2081 primitive Gaussians contracted to 1006 symmetry adapted basis functions. The other more important difference is that the spectrum (blue) of the complex is slightly red shifted from the N3 spectrum, which is due to a number of very weak charge transfer excitations between the dye and the nanocluster.

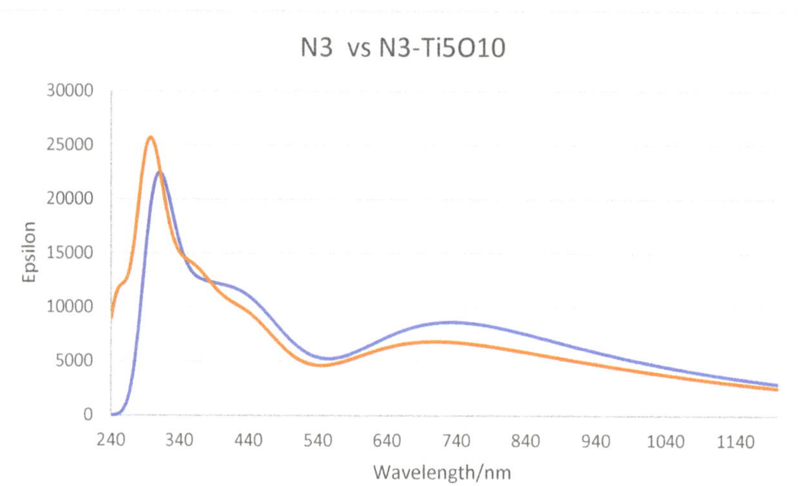

Figure 8. Absorption spectra of N3 dye (red) and the N3–$(TiO_2)_5^-$ complex (blue) simulated in vacuum. Full width at half maximum, FWHM broadening is 0.300 eV.

The above result is akin to experimental results where nearly identical absorption curves were found for N3 in DMSO and for colloidal N3 | TiO_2 in DMSO [23]. In addition, similar curves for N3 in ethanol solution and N3 adsorbed on nanocrystalline TiO_2 films were observed [21]. This observation that the low-energy bands of N3 and N719 in solution and on TiO_2 are very similar has been attributed to the similar roles that Ti^{4+} in the adsorbed state and protons in solution play in binding the oxygen atoms of carboxylate [21]. This conclusion is supported by our calculations, which show that the binding Ti atoms in our NP have a very high positive charge and are also the reason why the simulated N3 and N3–$(TiO_2)_5^-$ absorption spectra are so similar.

For N3 in ethanol solution, the MLCT bands in the experimental spectra were assigned at 398 nm and 538 nm in the visible, and bands at 314 nm and a shoulder at 304 nm in the UV were assigned to intra ligand π–π * transitions [21,60]. In our gas phase UV-VIS simulation of neutral N3 in Figure 8, we observe bands around 300, 370, 440, and 700 nm, which are quite different from the experimental spectrum. The largest difference is that the experimental peak at 538 nm appears to be shifted to 700 nm in the simulation. The TD-DFT calculations show this latter band is an MLCT excitation. Our three highest wavelength bands for N3 are very similar to those found by Kouki et al. in vacuum [57] at 373, 446, and 670 nm using B3LYP and a much larger basis set. In aqueous solution, the pH effect should be considered, which determines the charge on N3. At pH 11, the peaks are found at 308, 370, and 500 nm [5] presumably from the N3 dianion. We have also examined the effect on the gas phase absorption spectrum of deprotonation of N3. In Figure S1, the neutral N3, its singly charged anion, and the dianion (same as the dianion of N719) are compared. The peak for N3 at 300 nm is shifted to 320 nm in the N3 anion and 340 nm in the dianion; however, the peak in the visible at around 700 nm does not shift very much or vary systematically with charge. On the other hand, it is well known that theoretical simulations of Ru dyes, which include a solvent environment, give a much better representation of the experimental spectra [58,60,61]. In fact, if we simulate the spectrum of N3 in ethanol, as seen in Figure 9, we do find bands at 300, 378, 458, and 561 nm in the visible, which are much closer to experimental values. However, we could not use a solvent environment in our Raman simulations with the N3–$(TiO_2)_5^-$ complex, because in this case, the calculations would not converge. Therefore, we were forced to conduct the vibrational calculations in vacuum. Nevertheless, it should be noted that Raman simulations in vacuum show excellent agreement with experimental vibrational Raman frequencies in solution after an empirical scale factor correction. Thus, we use the TD-DFT vacuum excitation calculations to analyze the nature of the excited states that are used for exciting time-dependent Raman spectra and to analyze the CT process.

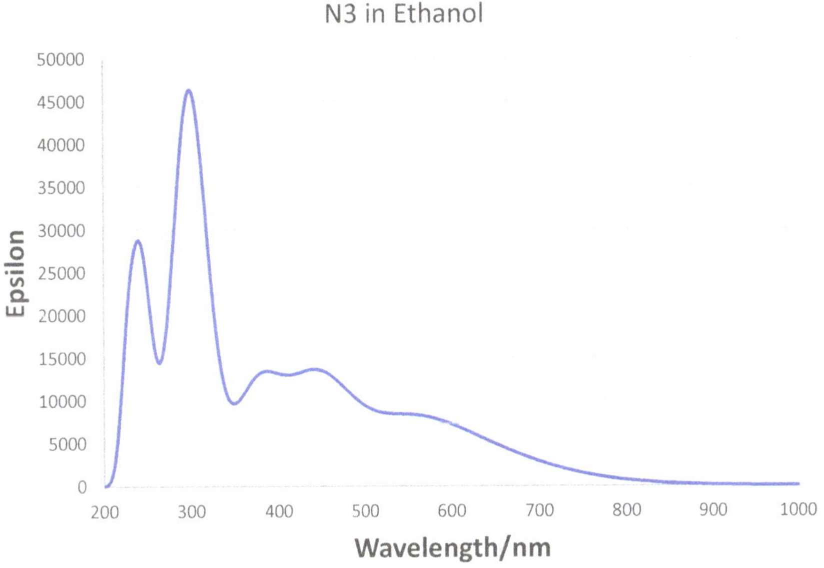

Figure 9. Absorption spectra of N3 dye simulated in ethanol using IEFPCM. FWHM broadening is 0.25 eV.

3.5. Excited State Transitions

Here, we report on the TD-DFT calculations of the optical transition of N3–$(TiO_2)_5^-$ in vacuum corresponding to the spectrum of the complex in Figure 8. We calculated 150 excited states of the complex; however, we will only be concerned with the first 45 excited states, which cover the near IR to around 400 nm. It turns out that the first 13 excited states cover the spectral region of 1200 to 648 nm and are all very similar to previous calculations for N3 [28]. These excitations have been described as MLCT transitions from mixed Ru isothiocyanate orbitals to the low-lying π^* orbitals of the dcbpy ligands. For our complex, the transitions described in the TD-DFT simulations all involve combinations of many delocalized molecular orbitals in both the hole and particle/electron parts of the excitations. Thus, in the first 10 excitations, all the hole states are composed of various combinations of HOMO, HOMO-1, HOMO-2, and HOMO-3, while the particle states involve various combinations of the LUMO, LUMO + 1, and LUMO + 2. These are in the gap below the putative conduction band edge; see Figure 7. We illustrate these transitions for representative excitations with the natural transition orbital, NTO, method of Martin [53] for hole and particle states. NTOs utilizes an orbital transformation that brings the transition density to diagonal form and allows a dramatic simplification in the qualitative description of electronic transitions.

The simplification is observed in Figure 10A where excited state 4 at 1.28 eV with oscillator strength f = 0.0151 is represented by five MO transitions, as shown in Table S1, with various coefficients in the TD-DFT output but by easily recognized orbitals in the hole/particle NTO iso-surfaces. This transition includes the MLCT of Ru-NCS hole to a particle with π^* dcbpy, which is not anchored and also to the anchored π^* pyridine and carboxylate part of the complex. In addition, it includes a ligand field transition from Ru $6d_{yz}$ to Ru $6d_z{}^2$. The excitation for excited state 8, Figure 10B, at 1.64 eV and f= 0.0624 is a stronger excitation, which involves seven MO transitions. Its hole/particle NTOs involves the transition of the Ru–$(NCS)_2$ hole to a particle with the non-anchored π * dcbpy. However, it also includes a ligand field transition of Ru $6d_z{}^2$ to Ru $6d_{yz}$. Both transitions in Figure 10 are below the energy of the LUMO + 3, which represents the conduction band edge and would not involve charge transfer to the TiO_2 part of the complex; however, they would support resonance Raman excitation.

All excitations for the first 45 excited states for the complex up to around 3.2 eV or 392 nm involve transitions from a hole state, which is either a Ru-NCS or Ru-$(NCS)_2$ NTO, and these are both illustrated in Figure 10. The first thirteen of these excited states are mainly excitations to π^* parts of the dcbpy ligands with some involvement of d orbitals of Ru. Starting at excited state 14, the excitation is a mixed transition with the particle state involving π^* dcbpy ligands, including the anchoring carboxylate and some $3d_Z{}^2$ orbitals on the Ti atoms; see Figure 11A. This excitation is at 2.0 eV with f = 0.0314. However, at excited state 17, the transition becomes a pure direct photoinduced charge transfer from Ru-NCS to the Ti $3d_Z{}^2$ orbital; see Figure 11B. This is similar to the direct photo-injection mechanism in the calculations of Persson et al. [35] for catechol on a $Ti_{38}O_{76}$ nanoparticle model. Furthermore, excited state 17 is more akin to excited state 18 at 1.98 eV of N719-$(TiO_2)_{82}$ complex of De Angelis et al. [33], which shows an electron state dominated by Ti(iV) t_{2g} orbitals with f = 0.03. For our model, excited state 17 at 2.13 eV has an oscillator strength of only f = 0.009, but it should support Herzberg–Teller scattering in a CT-SERS mechanism. These results indicate the possibility of a direct one-step CT mechanism for DSSCs.

The excited states were also examined with the charge transfer distance index, D_{CT}, which measures the spatial extent of charge transfer excitations in Å and the charge passed index, q_{CT}, which is the integration of the density depletion function over all space in atomic units, au. The index D_{CT} is calculated as the distance between the barycenters of density increment and density depletion regions. Table S2 shows the first hundred excited states with their values of D_{CT}, q_{CT}, and oscillator strengths. We arbitrarily assume that any state with a charge transfer distance index over 9.0 Å and charged passed index over

0.980 au is a charge transfer state. There are 10 such states in the first hundred excited states of the complex in Table S2, which have nonzero oscillator strength. For example, state 17 shown in Figure 11B has values of D_{CT} = 10.426 Å, q_{CT} = 0.985 au with an oscillator strength of 0.0009. State 21 which has hole/particle NTO iso-surfaces identical to those of excited state 17 has values of D_{CT} = 10.099 Å, q_{CT} = 0.982au with an oscillator strength of 0.0003. These 10 long-distance CT states have an oscillator strength average of only 0.0003, indicating little overlap between the ground and excited state wavefunctions; however, they all could support the CT-SERS mechanism by intensity borrowing.

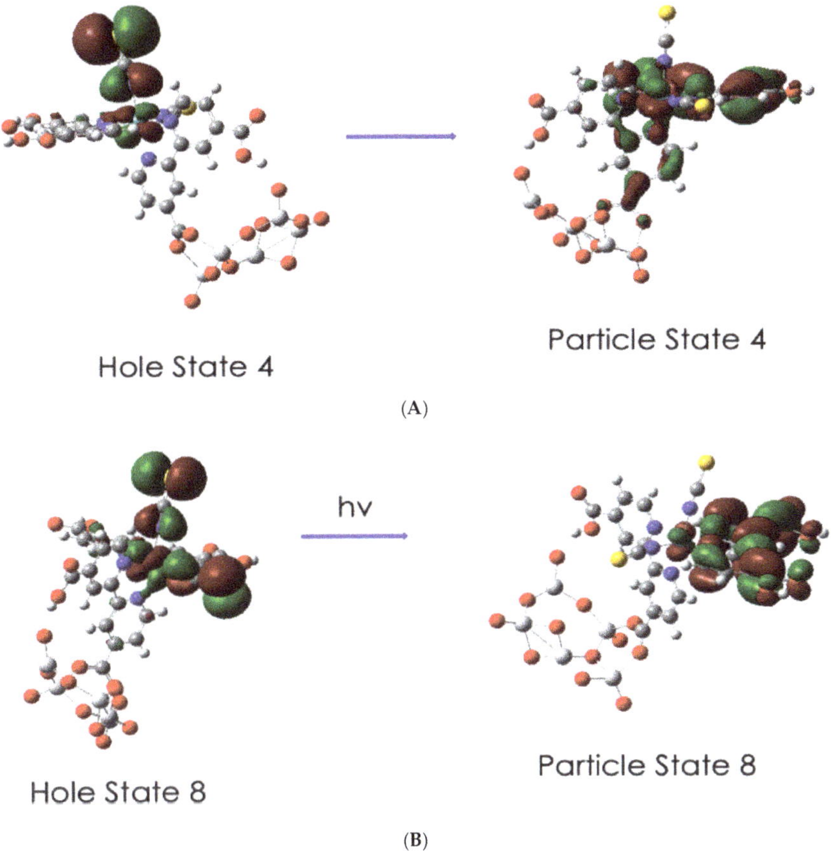

Figure 10. Hole and particle natural transition orbitals for excited states 4 and 8. (**A**). Excited state 4, energy = 1.2770 eV, f = 0.0151. (**B**). Excited state 8, energy = 1.6379 eV, f = 0.0624.

3.6. Simulated Raman Scattering Spectra

Figure 12 shows the normal Raman scattering (NRS) spectra of the two complexes A and B. Both are optimized structures of N3 anchored by a bridging bidentate carboxylate on Ti_5O_{10}, as shown in Figure 4, but complex B is tilted so that one of its isothiocyanate ligands is within 2.5 Å of the nanoparticle surface. The intensity of the NRS spectrum of complex B was doubled for clarity in Figure 12. For comparison, the strongest peak in both complexes at 1610 cm^{-1} is actually 2.3 more intense in complex A. The most striking feature of complex B is the two peaks around 2100 cm^{-1}. The stronger peak at 2165 cm^{-1} is the C = N vibration of the NCS ligand closest to the surface, while the peak at 2097 cm^{-1} is for the NCS ligand away from the surface. Only one peak at 2104 cm^{-1} shows up for complex A, which is the C = N vibration for both NCS ligands. There is also a very small peak for complex B in this region at 2122 cm^{-1} in between the more intense ones, which is the O-H vibration for –COOH forming a hydrogen bond to an O atom of Ti_5O_{10}. The effect of the surface on the C = N stretch is to split the vibration into two and to enhance the one for the NCS ligand closest to the surface. The band at 2165 cm^{-1} is close to the strong band at 2156 cm^{-1} for the pre-resonance Raman of N719 excited with 632.8 on silver colloid in acetonitrile where strong adsorption via the NCS groups was proposed [13]. A Ag_2@TiO_2 dimeric-shell–core NP also shows the band at 2155 cm^{-1} for N719, again indicating it is adsorbed via the two NCS groups [16]. This peak for N719 moves to 2104 cm^{-1} on TiO_2 surfaces [5,22,23] as we find with our simulation of complex A in which the two NCS groups are oriented away from the NP surface. As far as we know, experimental Raman spectra on TiO_2 surfaces have not shown two bands around 2100 cm^{-1}, which indicated that this tilted surface geometry of complex B is not prevalent.

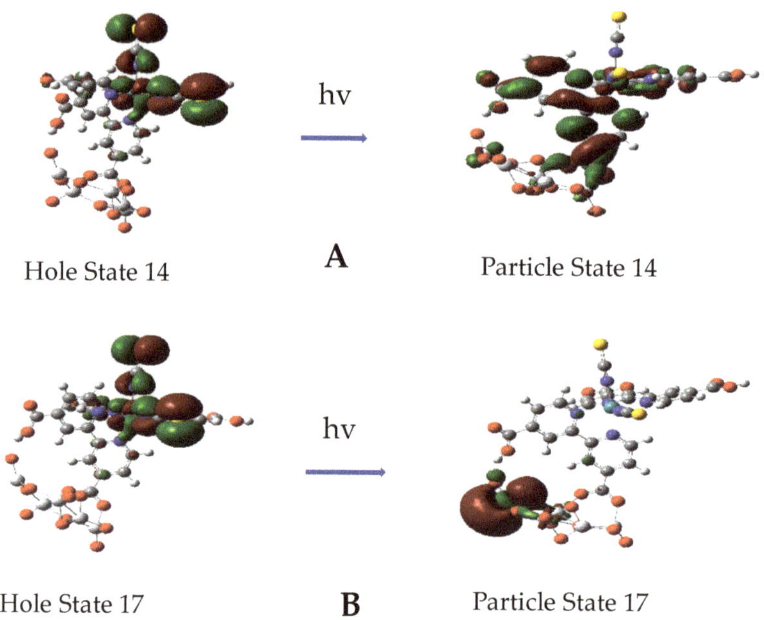

Figure 11. Hole and particle natural transition orbitals for excited states 14 and 17. (**A**): State 14, energy = 2.0049 eV, f = 0.0314. (**B**): State 17, energy = 2.1275 eV, f = 0.0009.

A comparison of the NRS for complex A and B shows that in the frequency range of 1000 to 1800 cm^{-1}, most major bands overlap with exceptions of a few weak bands such as the one at 1709 cm^{-1}. In Table 3, we compare our calculated NRS of complex A with four reports of experimental spectra in the literature. The experimental results are for N3 or N719 on various TiO_2 surfaces and in different solvents with reported excitation at 780, 514.5, 514, and at 415.5 nm. Considering the diversity of the experiments, the agreement of the Raman shift frequencies between the experiments and with our simulation is very good. This is reasonable, since bound N719 and N3 are alike because Ti(IV) supply positive charge to the anchoring groups, and the lack of symmetry shows the same frequencies even though excitation at 780 nm is NRS and 514 nm and 415 nm would give RRS. Important results revealed by the normal mode analysis of the N3-$(TiO_2)_5$ complex is that the band at 1534 cm^{-1} is not just a bipyridine ring stretch but contains O = C = O stretching of the anchoring group, which is in phase with the C = C stretching in the pyridine to which it is attached (see Figure S5). In addition, the band at 1388.5 cm^{-1} is shown to have the assignment of the symmetrical COO stretch for the carboxylate anchoring group on $(TiO_2)_5$ in the bridging bidentate structure, which is in-phase with C–H wags on its pyridine ring; see Figure S4. The assumption made in the literature has been that the band at 1388 cm^{-1} indicates either the bridging bidentate or the bidentate chelate binding [23,24]. Our result is evidence that this band is for the bridging bidentate structure. Shoute and Loppnow [23] indicate that this band was not observed in the Raman spectrum of isolated N3, and our simulation of the NRS spectrum of isolated N3 allows the same conclusion (Figure S2). Thus, it appears that the band at 1388 cm^{-1} indicates chemisorbed N3 (N719) in a bridging bidentate structure. In fact, there are three vibrational modes at 503, 1388, and 1534 cm^{-1} that contain vibrations of the anchoring carboxylate group. The displacement vectors of these normal modes are shown in Figures S3–S5. Furthermore, some of the low-frequency modes in Table 3 contain vibrations of the $(TiO_2)_5$ NP such as 836 and 796 cm^{-1}.

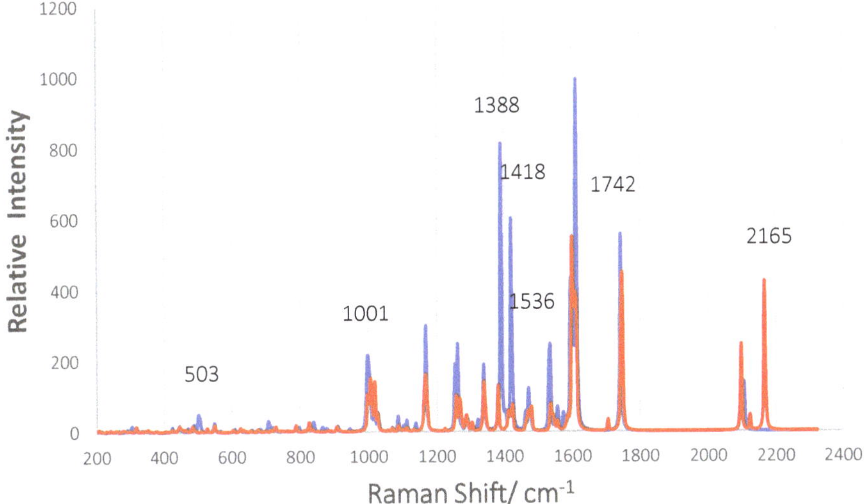

Figure 12. Normal Raman spectra of complex A (blue) and complex B (red). The intensity of complex B has been doubled for clarity. FWHM broadening 2 cm^{-1}. A scaling factor 0.970 have been used for all Raman frequencies.

Finally, we have simulated with TD-DFT the frequency-dependent Raman scattering spectra by exciting near two different excited states. These two excited states are S4 and S17, and their hole/particle NTOs are shown in Figures 10A and 11B, and their MO compositions are given in Table S1. Excited state S4 is an MLCT excitation at 1.277 eV (f = 0.0151), which is below the putative conduction band edge and should give resonance Raman scattering by a Franck–Condon scattering mechanism. Excited state S17 at 2.1275 eV (f = 0.009) is long distance charge transfer excitation (D_{CT} = 10.43 Å) right at the putative conduction band edge and should give Raman scattering by intensity borrowing from a nearby strong state through a Herzberg–Teller scattering mechanism [9].

Table 3. Comparison of our NRS results for complex A with the literature for N3 or N719 on a TiO_2 surface. A. TiO_2/N719 in 3MPN, 780 nm laser [22]. B. N3 | TiO_2 in DMSO, 514.5 nm laser [23]. C. N719-Aqueous TiO_2, 514 nm laser [25] D. N719 adsorbed on TiO_2, 415.44 nm laser [5].

A	B	C	D	This Paper	Assignments
2104	2104	2095	2105	2104	(N = C) stretch in NCS
1735		1727	1726	1742	(C = O) stretch in COOH
1611	1610	1605	1610	1613	bpy ring stretch in anchor dcbyp
1544	1541	1542	1539	1534	bpy ring stretch and anchor. O = C=O stretch
	1469	1474	1471	1471	ring stretch in both bpy ligands
1433			1430	1419	i.p. C–H wag on both bpy, C–CO stretch
	1388	1376	1367	1388	i.p. C–H wag on anchored bpy, sym. stretch of anchored COO
1303		1315	1331	1338	i.p. C-H wag on unbound dcbpy, O-H bend on both COOH of dcbpy
1290	1294			1288	i.p. C-H wag on unbound dcbpy, O-H bend on both COOH of this dcbpy
1260	1256	1268	1260	1261	unsym. ring stretch on unbound dcbpy, O-H bend on one COOH
			1222	1252	i.p. C–H wag on unbound dcbpy
	1130		1155	1167	i.p. C–H wag on unbound dcbpy, O-H bend on both COOH of dcbpy
1106	1102	1106	1102	1111	sym. i.p. C-H wag on unbound dcbpy
	1021	1020		1001	Trigonal ring stretch on both bpy
	920			839	o. p. C–H wag on anchored bpy, Ti–O–Ti stretch,
	809			827	C = S stretch, dcbpy deformation, anchored COO bend
702	750			796	unsym. Ti-O-Ti stretch
	698	699		719	unsym. ring stretch on anchored bpy, O–H bend in COOH H-bond to surface
	454	512		503	Ti–O = C stretch of anchored COO grps, o.p. bpy ring deformation.
	364	397		376	Ru-NCS wag, unbound dcbpy rock
	318			320	Ru-N(bpy)$_{trans}$ wag

The TD-DFT simulations are dynamic polarizability calculations that follow the outline of Neugebauer et al. [62] and are based on the Placzek theory with the double harmonic approximation (harmonic oscillator force field and truncation of the Taylor expansion of the polarizability with respect to a normal mode after the quadratic term). This requires numerical derivatives of the polarizability tensor components with respect to a normal coordinate. The dynamic polarizability is found at each band position by calculating the frequency-dependent polarizability using pre-resonance Raman code [50]. If we formulate the dynamic polarizability tensor components in terms of the Kramers–Heisenberg–Dirac (KDH) expression through second-order perturbation theory [63], using the Born–Oppenheimer approximation, expanding the electronic transition moments, μ, in a Taylor expansion to first power in the normal coordinates Q, and utilizing the Herzberg–Teller, HT, formulation for mixing of vibronic functions, we obtain Equations (2)–(4)with the following familiar A and B terms of Albrecht [64] as we developed [63]

$$\alpha_{\sigma\rho} = \mathrm{A} + \mathrm{B} \tag{2}$$

$$A = \sum_{S=F,K\neq I} \sum_{k} \left[\frac{\mu^{\sigma}_{SI}\mu^{\rho}_{SI}}{\hbar(\omega_{SI} - \omega)} + \frac{\mu^{\rho}_{SI}\mu^{\varrho}_{SI}}{\hbar(\omega_{SI} + \omega)} \right] \langle i|k >< k|f\rangle, \tag{3}$$

$$\begin{aligned} B = & \sum_{R=F,K} \sum_{S=F,K} \left[\frac{\mu^{\sigma}_{IR} h_{RS} \mu^{\rho}_{SI}}{\hbar(\omega_{RI} - \omega)} + \frac{\mu^{\rho}_{IR} h_{RS} \mu^{\rho\varrho}_{SI}}{\hbar(\omega_{RI} + \omega)} \right] \frac{\langle i|k>\langle k|Q_k|f\rangle}{\hbar\omega_{RS}} + \\ & \sum_{R=F,K} \sum_{S=F,K} \sum_{k} \left[\frac{\mu^{\sigma}_{IS} h_{SR} \mu^{\rho}_{RI}}{\hbar(\omega_{RI} - \omega)} + \frac{\mu^{\rho}_{IR} h_{RS} \mu^{\rho\varrho}_{SI}}{\hbar(\omega_{RI} + \omega)} \right] \frac{\langle i|Q_k|k\rangle < k|f\rangle}{\hbar\omega_{RS}} \end{aligned} \tag{4}$$

where $E_{SI} = \hbar\omega_{SI} = E_S - E_I$, the energy between the ground electronic state I and an excited electronic state S, etc., the exciting light is $h\omega$, $<i\,|$, and $|\,f>$ are initial and final vibrational wavefunctions of the ground state, $<k\,|$ is the wavefunction of an intermediate state, and σ, ρ are the polarization directions. When there is a resonance, the second term in the square brackets of each summation is negligible with respect to the first term. In these equations, we have not included the imaginary bandwidth, $i\gamma_k$, in the resonance denominator. The A-term is the Franck–Condon term responsible for molecular resonance Raman, and the B-term is Herzberg–Teller term responsible for a long-range charge-transfer resonance between molecule and nanoparticle, CT-SERS. In these expressions, the sums range over all excited states (R and S), which include both charge transfer states and molecule or nanoparticle states (but of course exclude terms for which a denominator vanishes (such as

S or $R = I$)). In order to examine the situation in which the light is not only resonant with the molecular excitations but also a charge-transfer transition in the nanoparticle–molecule system, we select for examination only terms in the general expression (1) above that include a charge-transfer state. These terms contain the linear Herzberg–Teller vibronic coupling constant h_{SR}

$$h_{SR} = \left\langle S_e, 0 \middle| \left(\frac{\partial V_{eN}}{\partial Q} \right)_0 \middle| R_e, 0 \right\rangle \tag{5}$$

where V_{eN} is the electron–nuclear attraction term in the Hamiltonian, which is evaluated at the equilibrium nuclear positions (0), and the purely electronic transition moment between states are written: $\mu^{\sigma}{}_{SI} = <S_e \mid \mu^{\sigma} \mid I_e>$, $\mu^{\sigma}{}_{RI} = <R_e \mid \mu^{\sigma} \mid I_e>$ and $\mu^{\sigma}{}_{SR} = <S_e \mid \mu^{\sigma} \mid R_e>$. Since in long-range charge transfer, the dipole moment μ_{SI} of Equation (2) is very small, it is the B term in Equation (3) above that can dominate the A term and give rise to charge transfer Raman scattering when there is coupling between electronic excited states R and S with finite Herzberg–Teller vibronic coupling constant [65].

The simulated pre-resonance Raman spectra for excitation near excited state 4, which is a MLCT transition, and near excited state 17, which is long-distance dye-to-TiO_2 CT transition, are illustrated in Figure 13. The excitation for the former spectrum (blue) at 955.3 nm is 169 cm^{-1} above excited state 4, while the excitation for the latter spectrum (orange) is 130 cm^{-1} below excited state 17. The vertical axis for state 4, the MLCT transition, is on the left of Figure 13 and is a factor of three higher than the vertical axis for state 17, the N3-to-$(TiO_2)_5$ CT transition, on the right of the figure. The MLCT pre-resonance Raman spectrum is about 6.5 times more intense than the N3-to-$(TiO_2)_5$ CT pre-resonance Raman spectrum. On the other hand, the ratio of oscillator strength is about 17 higher for the MLCT process compared to the N3-to-TiO_2 CT process. In fact, the pre-resonance spectrum for excited state 17 and for excited state 4 are about 2×10^2 and about 1.3×10^3 times the simulated normal Raman spectrum, respectively. The fact that the oscillator strength for excited state 17 is so low and yet there is significant enhancement of the pre-resonance Raman spectrum indicates the intensity borrowing mechanism of Equation (3) is operating.

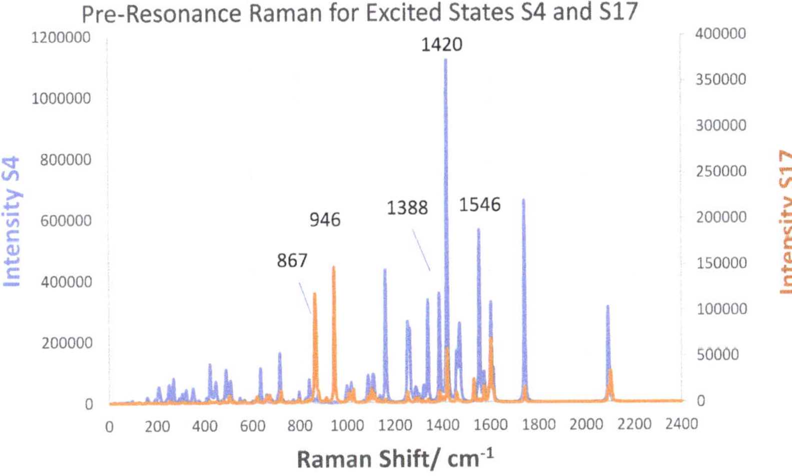

Figure 13. Pre-resonance Raman spectra of N3–$(TiO_2)_5$ for complex A. Blue spectrum excited near S4, $f = 0.0151$ at 955.27 nm (10,468.9 cm^{-1}). Orange spectrum excited near state S17, $f = 0.0009$ at 587.20 nm (17,029.9 cm^{-1}) FWHM broadening 2 cm^{-1}. A scaling factor 0.970 has been used for the Raman frequencies.

There are some bands that become more intense in these pre-resonance Raman spectra. The two strongest bands in the N3-to-$(TiO_2)_5$ CT spectrum are at 867 and 946 cm^{-1}, and both involve the $(TiO_2)_5$ NP. They are a Ti–O stretching vibration coupled to the O–H bend of the carboxylate group hydrogen bonded to the vibrating Ti–O group (946 cm^{-1}) and an out-of-plane C–H wag in the pyridine-ring whose carboxylate is H-bonded to the Ti–O (867 cm^{-1}). These C–H wags are also coupled to the Ti–O–Ti stretching vibration in the nanoparticle. The group of bands in the MLCT pre-resonance Raman spectrum at 1611, 1573, and 1472 cm^{-1} have the same intensity ratios as experimental MLCT resonance Raman spectra of N3 or N719 on titania in the literature [5,23,25]. In addition, the 1388 cm^{-1} for the symmetrical COO stretching vibration is still observable in both pre-resonance spectra. A noticeable difference in the MLCT RRS is the band at 1420 cm^{-1}, which becomes the most intense band in the spectrum. This vibration (Figure S6) contains in-plane C–H wags on all the bipyridine ligands but also has a contribution from the symmetrical O = C=O stretch of the anchoring carboxylate group. It is somewhat similar to the normal mode at 1388.5 cm^{-1}. In fact, there is a weak band in the resonance Raman spectrum excited at 414.44 nm of Nazeeruddin et al. [5] at 1430 cm^{-1} and a band reported at 1433 cm^{-1} reported by Greijer at al. [22] which may correspond to this band. Since the excited states of the N3–$(TiO_2)_5^{1-}$ complex, which is an anion and a very small model NP system, will not correspond to those of N3 or N719 adsorbed on a much larger TiO_2 surface, which should have a continuum of conduction band states, it is difficult to correlate band intensities of these simulated pre-resonance spectra with experimental frequency dependent spectra. However, the normal modes that we have determined with the simulation of these spectra should certainly be valid and of interest for future experimental studies.

4. Conclusions

The geometrical structure of the Ruthenium dicarboxylbipyridyl dye N3 (whether in its isolated neutral and anionic forms or as an adsorbate bound to a $(TiO_2)_5$ NP) was found by DFT calculations to have the same distorted octahedral structure, which splits the d orbital manifold of Ru. The bound complex had two configurations, both which anchor N3 through a carboxylate bridging bidentate geometry to two Ti atoms and also involve a hydrogen bond formed by the other –COOH group on the binding bipyridine to a dangling O atom of the NP. One form of the complex has the isothiocyanate groups directed away from the surface, while a more tightly bound form, as indicated by binding energy calculations, has only one of the isothiocyanate groups interacting with the surface. This form of the surface geometry can be detected, since it has two N = C stretching vibrations from the NCS groups at 2104 and 2165 cm^{-1}, whereas the other form with NCS groups directed away from the surface shows only the 2104 cm^{-1} stretching vibration. The strong interaction of NCS groups from N3 or N719 is more likely to be observed on metals like Au or Hg and has been detected on the Hg and Au [12,13,17,18] and on Ag@TiO_2 and Ag_2@TiO_2 [15,16], where in all these cases, only the higher wavenumber vibration shows up in the spectra, indicating both NCS groups binding strongly to the surface.

The simulations also show that the band at 1388 cm^{-1}, which contains the symmetrical stretching vibration of the bridging bidentate carboxylate group and which does not show up in the isolated dye, clearly indicates chemisorption of the bound dye with this surface geometry, as deduced experimentally from bands close to 1380 cm^{-1} [18,19,23]. For most of the bands, the DFT calculations show that the normal Raman spectrum of N3 adsorbed on TiO_2 is similar to isolated N3 because of the positive charge on the Ti atoms. These form a chemical bond with the two oxygen atoms of the carboxylate of the dye similar to hydrogen ions in solution. This conclusion was previously deduced from an experimental optical absorption investigation in the literature [21]. Thus, our DFT structure and Raman calculations confirm many of the explanations inferred from experimental studies. On the other hand, the simulated pre-resonance spectra of both types of resonance mechanisms show enhancement of vibrational modes involving combined dye and TiO_2 vibrations, such as those at 867, 946, and 1420 cm^{-1}.

The electronic structure calculations made for N3–$(TiO_2)_5$ agree with similar calculations made for much larger dye-$(TiO_2)_n$ models (n = 38 or 82) with respect to molecular orbital iso-surfaces and putative band edge gaps. Furthermore, the optimized structural results agree with experimental crystal X-ray average structure data, and thus, N3-$(TiO_2)_5{}^{1-}$ proved to be an adequate model for a single bridging bidentate surface geometry with an additional hydrogen-bonded carboxylate to the NP surface. This relatively small structural model allows time-dependent TD-DFT Raman simulations with a B3LYP/6-31 + G(d)/LANL2DZ calculational model and shows that both two-step intermediate and one-step direct charge transfer mechanisms are possible with N3 dye on TiO_2. The long-distance dye-to-NP CT is readily determined to be appropriate for all excited states with charge transfer distance index, D_{CT}, over 9.0 Å and a charged passed index, q_{CT}, over 0.980 au. The optimized geometry with the anchoring group in our model shows Ti-O bond distances of 2.02 Å and 2.06 Å in the C-O-Ti arms of the bridge, indicating a reasonably strong chemical bond anchoring N3 to the surface. The molecular orbital analysis of the N3-$(TiO_2)_5$ complex such as the LUMO + 3 level (Figure 6B) shows delocalization across the anchoring bridge, indicating strong electron coupling facilitating ultrafast electron injection. The Herzberg–Teller resonance Raman scattering theory for long-distance photoinduced CT has implications for photoinduced electron injection at dye-covered TiO_2 surfaces. Vibronic coupling CT theory analogous to Herzberg–Teller theory has been used to analyze charge-transfer states in organic solar cells [66]. This theory adopts a three-state model involving a ground state photoexcited to a donor-acceptor state, D^+A^-, which is vibronically coupled to a local, strongly absorbing excited state on the donor or acceptor.

The dynamic polarizability expression shows that both Franck–Condon and Herzberg–Teller resonance Raman scattering mechanisms are possible depending on transition dipole moments and their derivatives. The MLCT and long-distance dye-to-NP charge-transfer (CT-SERS) pre-resonance Raman scattering mechanisms for the two excited states investigated result in enhancement factors of around 1×10^3 and 2×10^2, respectively. These simulated pre-resonance MLCT and CT-SERS enhancement factors predict many orders of magnitude lower in intensity compared with SERS experiments on Ag, Au, and Ag core–(TiO_2) shell SERS substrates. However, the enhancement factors are representative for these two types of mechanisms on TiO_2 surfaces, and the Raman intensities in an experimental case will depend on the oscillator strength of excited states and excited life-time broadening mechanisms. In fact, the optical simulations show many MLCT excited states with two to four times the oscillator strength of the representative excited state 4. Our results suggest that experimental Raman spectra excited at different excitation energies should be able to distinguish between these types of mechanisms on titania surfaces. However, it would be important to extend these studies in future simulations of dye–TiO_2 Raman spectra to more accurate models. These extensions should include a model with N3 or other dyes adsorbed on a TiO_2 structure, which allows a quasi-continuum of TiO_2 conduction band states, implementation of a solvent environment, and modelling of the effect of surface protonation on surface geometry and thus on Raman spectra.

Simple Summary: This study verifies that a small sized model of the semiconductor TiO_2 can be used to accurately study the properties of Ru bipyridyl dye, called N3, when the dye is adsorbed on its surface. This system is a model of a photoelectrochemical device for conversion of solar energy to electric current called a Dye-Sensitized Solar Cell (DSSC). Various types of Raman spectra were simulated which allow the surface geometry of the dye to be predicted. Also, the mechanisms of how the light photon causes an electron to be injected from the dye into semiconductor surface were investigated by quantum mechanical electronic structure calculations.

Supplementary Materials: The following are available online at https://www.mdpi.com/article/10.3390/nano11061491/s1. Optical spectra (Figure S1) and normal Raman spectra (Figure S2). Composition of selected excited states (Table S1), charge transfer distance, charge passed indices and oscillator strengths (Table S2), and normal mode vector displacement diagrams of vibrational frequencies (Figures S3–S6).

Author Contributions: R.L.B. and J.R.L. conceived, reviewed, and edited the manuscript. R.L.B. did the formal analysis and data curation. All authors have read and agreed to the published version of the manuscript.

Funding: We are indebted to the National Science Foundation (CHE-1402750) for partial funding of this project. This work was also partially supported by NSF grant number HRD-1547830 (IDEALS CREST). Computer facilities for this research were supported by a XSEDE Grant CHE090043 and by the CUNY High Performance Computer Center. This work used the Extreme Science and Engineering Discovery Environment (XSEDE), which is supported by National Science Foundation grant number ACI-1053575.

Institutional Review Board Statement: Not applicable.

Informed Consent Statement: Not applicable.

Data Availability Statement: The data available are in Supplementary Materials.

Conflicts of Interest: The authors declare no conflict of interest.

References

1. O'Regan, B.; Grätzel, A. Low-Cost, High-Efficiency Solar Cell Based on Dye-Sensitized Colloidal TiO_2 Films. *Nature* **1991**, *353*, 737–740.
2. Bai, Y.; Mora-Sero', I.; De Angelis, F.; Biisquert, J.; Wang, P. Titanium Dioxide Nanomaterials for Photovoltaic Applications. *Chem. Rev.* **2014**, *114*, 10095–10130. [PubMed]
3. De Angelis, F.; Fantacci, S.; Selloni, A.; Grätzel, M.; Nazeeruddin, M.K. Influence of the Sensitizer Adsorption Mode on the Open-Circuit Potential of Dye-Sensitized Solar Cells. *Nano Lett.* **2007**, *7*, 3189–3195. [CrossRef]
4. Lee, C.-P.; Li, C.-T.; Ho, K.-C. Use of organic materials in dye-sensitized solar cells. *Mater. Today* **2017**, *20*, 267–283. [CrossRef]
5. Nazeeruddin, M.K.; Zakeeruddin, S.M.; Humphry-Baker, R.; Jirousek, M.; Liska, P.; Vlachopoulos, N.; Shklover, V.; Fischer, C.-H. GrätzeAcid-Base Equilibria of (2,2′-Bipyridyl-4,4′-dicarboxylic acid) ruthenium(II) Complexes and the Effect of Protonation on Charge-Transfer Sensitization of Nanocrystalline Titania. *Inorg. Chem.* **1999**, *38*, 6298–6305. [CrossRef]
6. Yang, L.; Jiang, X.; Ruan, W.; Zhao, B.; Xu, W.; Lombardi, J.R. Observation of Enhance Raman Scattering for Molecules Adsorbed on TiO_2 Nanoparticles: Charge-Transfer Contribution. *J. Phys. Chem. C* **2008**, *112*, 20095–20098. [CrossRef]
7. Musumeci, A.; Gosztola, D.; Schiller, T.; Dimitrijevic, N.M.; Mujica, V.; Martin, D.; Rajh, T. SERS of Semiconducting Nanoparticles (TiO_2 Hybrid Composites). *J. Am. Chem. Soc.* **2009**, *131*, 6040–6041. [CrossRef]
8. Lombardi, J.R.; Birke, R.L. Theory of Surface-Enhanced Raman Scattering in Semiconductors. *J. Phys. Chem. C* **2014**, *118*, 11120–11130. [CrossRef]
9. Lombardi, J.; Birke, R.L. A Unified View of Surface-Enhanced Raman Scattering. *Acc. Chem. Res.* **2009**, *42*, 734–742. [CrossRef]
10. Wang, X.; Shi, W.; Wang, S.; Zhao, H.; Lin, J.; Yang, Z.; Chen, M.; Guo, L. Two-Dimensional Amorphous TiO_2 Nanosheets Enabling High-Efficiency Photoinduced Charge Transfer for Excellent SERS Activity. *J. Am. Chem. Soc.* **2019**, *141*, 5856–5862. [CrossRef]
11. Maznichenko, D.; Venkatakrishnan, K.; Tan, B. Stimulating Multiple SERS Mechanisms by a Nanofibrous Three-Dimensional Network Structure of Titanium Dioxide (TiO_2). *J. Phys. Chem. C* **2013**, *117*, 578–583. [CrossRef]
12. Theil, F.; Zedler, L.; März, A.; Xie, W.; Csáki, A.; Fritzsche, W.; Cialla, D.; Schmitt, M.; Popp, J.; Dietzek, B. Evidence for SERRS Enhancement in the Spectra of Ruthenium Dye−Metal Nanoparticle Conjugates. *J. Phys. Chem. C* **2013**, *117*, 1121–1129. [CrossRef]
13. Perez León, C.; Kador, L.; Peng, B.; Thelakkat, M. Influence of the Solvent on Ruthenium(II) Bipyridyl Complexes. *J. Phys. Chem. B* **2005**, *109*, 5783–5789. [CrossRef] [PubMed]
14. Schade, L.; Franzka, S.; Biener, M.; Biener, J.; Hartmann, N. Surface-enhanced Raman spectroscopy on laser-engineered ruthenium dye-functionalized nanoporous gold. *App. Surf. Sci.* **2013**, *347*, 19–22. [CrossRef]
15. Qiu, Z.; Zhang, M.; Wu, D.-Y.; Ding, S.-Y.; Zuo, Q.-Q.; Huang, Y.-F.; Shen, W.; Lin, X.-D.; Tian, Z.-Q.; Mao, B.-W. Raman Spectroscopic Investigation on TiO_2–N719 Dye Interfaces Using Ag@TiO_2 Nanoparticles and Potential Correlation Strategies. *Chem. Phys. Chem.* **2013**, *14*, 2217–2224. [CrossRef] [PubMed]
16. Zuo, Q.-Q.; Feng, Y.-L.; Chen, S.; Qiu, Z.; Xie, L.-Q.; Xiao, Z.-Y.; Yang, Z.-L.; Mao, B.-W.; Tian, Z.-Q. Dimeric Core−Shell Ag2@TiO_2 Nanoparticles for Off-Resonance Raman Study of the TiO_2−N719 Interface. *J. Phys. Chem. C* **2015**, *119*, 18396–18403. [CrossRef]
17. Wang, X.; Wang, Y.; Sui, H.; Zhang, X.; Su, H.; Cheng, W.; Han, X.-X.; Zhao, B. Investigation of Charge Transfer in Ag/N719/TiO_2 Interface by Surface-Enhanced Raman Spectroscopy. *J. Phys. Chem. C* **2016**, *120*, 13078–13086. [CrossRef]
18. Wang, X.; Zhao, B.; Li, P.; Han, X.X.; Ozaki, Y. Charge Transfer at the TiO_2/N3/Ag Interface Monitored by Surface-Enhanced Raman Spectroscopy. *J. Phys. Chem. C* **2017**, *121*, 5145–5153. [CrossRef]
19. Finnie, K.S.; Bartlett, J.R.; Woolfrey, J.L. Vibrational Spectroscopic Study of the Coordination of (2,2¢-Bipyridyl-4,4′-dicarboxylic acid) ruthenium(II) Complexes to the Surface of Nanocrystalline Titania. *Langmuir* **1998**, *14*, 2744–2749. [CrossRef]
20. Deacon, G.B.; Phillips, R.J. Relationships between the carbon-oxygen stretching frequencies of carboxylato complexes and the type of carboxylate coordination. *Coord. Chem. Rev.* **1980**, *33*, 227–250. [CrossRef]

21. Nazeeruddin, M.K.; Humphry-Baker, R.; Liska, P.; Grätzel, M. Investigation of Sensitizer Adsorption and the Influence of Protons on Current and Voltage of a Dye-Sensitized Nanocrystalline TiO_2 Solar Cell. *J. Phys. Chem. B* **2003**, *107*, 8981–8987. [CrossRef]
22. Greijer, H.; Lindgren, J.; Hagfeldt, A. Resonance Raman Scattering of a Dye-Sensitized Solar Cell: Mechanism of Thiocyanato Ligand Exchange. *J. Phys. Chem. B* **2001**, *105*, 6314–6320. [CrossRef]
23. Shoute, L.C.T.; Loppnow, G.R. Excited-State Metal-to-Ligand Charge Transfer Dynamics of a Ruthenium(II) Dye in Solution and Adsorbed on TiO_2 Nanoparticles from Resonance Raman Spectroscopy. *J. Am. Chem. Soc.* **2003**, *125*, 15636–15646. [CrossRef]
24. Perez León, C.; Kador, L.; Peng, B.; Thelakkat, M. Characterization of the Adsorption of Ru-bpy Dyes on Mesoporous TiO_2 Films with UV-Vis, Raman, and FTIR Spectroscopies. *J. Phys. Chem. B* **2006**, *110*, 8723–8730. [CrossRef]
25. Lee, K.E.; Gomez, M.A.; Elouatik, S. Further Understanding of the Adsorption Mechanism of N719 Sensitizer on Anatase TiO_2 Films for DSSC Applications Using Vibrational Spectroscopy and Confocal Raman Imaging. *Langmuir* **2010**, *26*, 9575–9583. [CrossRef]
26. Schiffmann, F.; VandeVondele, J.; Hutter, J.; Wirz, R.; Urakawa, A.; Baiker, A. Protonation-Dependent Binding of Ruthenium Bipyridyl Complexes to the Anatase (101) Surface. *J. Phys. Chem. C* **2010**, *114*, 8398–8404. [CrossRef]
27. Shklover, V.; Ovchinnikov, Y.E.; Braginsky, L.S.; Zakeeruddin, S.M.; Grätzel, M. Structure of Organic/Inorganic Interface in Assembled Materials Comprising Molecular Components. Crystal Structure of the Sensitizer Bis [(4,4′-carboxy-2,2¢-bipyridine)(thiocyanato)] ruthenium(II). *Chem. Mater.* **1998**, *10*, 2533–2541. [CrossRef]
28. Fantacci, S.; De Angelis, F.; Selloni, A. Absorption Spectrum and Solvatochromism of the [Ru(4,4-COOH-2,2-bpy)2(NCS)2] Molecular Dye by Time Dependent Density Functional Theory. *J. Am. Chem. Soc.* **2003**, *125*, 4381–4387. [CrossRef] [PubMed]
29. Nazeeruddin, M.K.; De Angelis, F.; Fantacci, S.; Selloni, A.; Viscardi, G.; Liska, P.; Ito, S.; Takeru, B.; Grätzel, M. Combined Experimental and DFT-TDDFT Computational Study of Photoelectrochemical Cell Ruthenium Sensitizers. *J. Am. Chem. Soc.* **2005**, *127*, 16835–16847. [CrossRef]
30. Persson, P.; Lundqvist, M.J. Calculated Structural and Electronic Interactions of the Ruthenium Dye N3 with a Titanium Dioxide Nanocrystal. *J. Phys. Chem. B* **2005**, *109*, 11918–11924. [CrossRef]
31. De Angelis, F.; Fantacci, S.; Selloni, A.; Nazeeruddin, M.K.; Grätzel, M. Time-Dependent Density Functional Theory Investigations on the Excited States of Ru(II)-Dye-Sensitized TiO_2 Nanoparticles: The Role of Sensitizer Protonation. *J. Am. Chem. Soc.* **2007**, *129*, 14156–114157. [CrossRef]
32. De Angelis, F.; Fantacci, S.; Selloni, A.; Nazeeruddin, M.K.; Grätzel, M. First-Principles Modeling of the Adsorption Geometry and Electronic Structure of Ru(II) Dyes on Extended TiO_2 Substrates for Dye-Sensitized Solar Cell Applications. *J. Phys. Chem. C* **2010**, *114*, 6054–6061. [CrossRef]
33. De Angelis, F.; Fantacci, S.; Moscani, E.; Nazeeruddin, M.K.; Grätzel, M. Absorption Spectra and Excited State Energy Levels of the N719 Dye on TiO_2 in Dye-Sensitized Solar Cell Models. *J. Phys. Chem. C* **2011**, *115*, 8825–88316. [CrossRef]
34. Prezhdo, O.V.; Duncan, W.R.; Prezhdo, V.V. Dynamics of the Photoexcited Electron at the Chromophore–Semiconductor Interface. *Acc. Chem. Res.* **2008**, *41*, 339–348. [CrossRef]
35. Persson, P.; Bergström, R.; Lunell, J.S. Quantum Chemical Study of Photoinjection Processes in Dye-Sensitized TiO_2 Nanoparticles. *J. Phys. Chem. B* **2000**, *104*, 10348–10351. [CrossRef]
36. Tae, E.L.; Lee, S.H.; Lee, J.K.; Yoo, S.S.; Kang, J.E.; Yoon, K.B. A Strategy to Increase the Efficiency of the Dye-Sensitized TiO_2 Solar Cells Operated by Photoexcitation of Dye-to-TiO_2 Charge-Transfer Bands. *J. Phys. Chem. B* **2005**, *109*, 22513–22522. [CrossRef] [PubMed]
37. Duncan, W.R.; Stier, W.M.; Prezhdo, O.V. Ab Initio Nonadiabatic Molecular Dynamics of the Ultrafast Electron Injection across the Alizarin-TiO_2 Interface. *J. Am. Chem. Soc.* **2005**, *127*, 7941–7951. [CrossRef]
38. Hannappel, T.; Burfeindt, B.; Storck, W.; Willig, F. Measurement of Ultrafast Photoinduced Electron Transfer from Chemically Anchored Ru-Dye Molecules into Empty Electronic States in a Colloidal Anatase TiO_2 Film. *J. Phys. Chem. B* **1997**, *101*, 6799–6802. [CrossRef]
39. Walters, A.; Gaal, D.A.; Hupp, J.T. Interfacial Charge Transfer and Colloidal Semiconductor Dye-Sensitization: Mechanism Assessment via Stark Emission Spectroscopy, *J. Phys. Chem. B.* **2002**, *106*, 5139–5142. [CrossRef]
40. Zhao, Y.; Liang, W.-Z. Theoretical investigation of resonance Raman scattering of dye molecules absorbed on semiconductor surfaces, *J. Chem. Phys.* **2011**, *135*, 044108. [CrossRef]
41. Ramakrishna, G.; Ghosh, H.N. Emission from the Charge Transfer State of Xanthene Dye-Sensitized TiO_2 Nanoparticles: A New Approach to Determining Back Electron Transfer Rate and Verifying the Marcus Inverted Regime. *J. Phys. Chem. B* **2001**, *105*, 7000–7008. [CrossRef]
42. Asbury, J.B.; Hao, E.; Wang, Y.; Ghosh, H.N.; Lian, T. Ultrafast Electron Transfer Dynamics from Molecular Adsorbates to Semiconductor Nanocrystalline Thin Films. *J. Phys. Chem. B* **2001**, *105*, 4545–4557. [CrossRef]
43. Wang, H.-F.; Chen, L.-Y.; Su, W.-N.; Chung, J.-C.; Hwang, B.-J. Effect of the Compact TiO_2 Layer on Charge Transfer between N3 Dyes and TiO_2 Investigated by Raman Spectroscopy. *J. Phys. Chem. C* **2010**, *114*, 3185–3189. [CrossRef]
44. Ma, S.; Livingstone, R.; Zhao, B.; Lombardi, J.R. Enhanced Raman Spectroscopy of Nanostructured Semiconductor Phonon Modes. *J. Phys. Chem. Lett.* **2011**, *2*, 671–674. [CrossRef]
45. Nomoto, T.; Fujio, K.; Sasahara, A.; Okajima, H.; Koide, N.; Katayama, H.; Onishi, H. Phonon mode of TiO_2 coupled with the electron transfer from N3 dye. *J. Chem. Phys.* **2013**, *138*, 224704. [PubMed]

46. Lundqvist, M.J.; Nilsing, M.; Persson, P.; Lunell, S. DFT Study of Bare and Dye-SensitizedTiO_2 Clusters and Nanocrystals. *Int. J. Quantum Chem.* **2006**, *106*, 3214–3234. [CrossRef]
47. Oprea, C.I.; Gîrtu, M.A. Structure and Electronic Properties of TiO_2 Nanoclusters and Dye-Nanocluster Systems Appropriate to Model Hybrid Photovoltaic or Photocatalytic Applications. *Nanomaterials* **2019**, *9*, 357. [CrossRef]
48. Berardo, E.; Hu, H.-S.; Shevlin, S.A.; Woodley, S.M.; Kowalski, K.; Zwijnenburg, M.A. Modeling Excited States in TiO_2 Nanoparticles: On the Accuracy of a TD-DFT Based Description. *J. Chem. Theory Comput.* **2014**, *10*, 1189–1199. [CrossRef] [PubMed]
49. Frisch, M.J.; Trucks, G.W.; Schlegel, H.B.; Scuseria, G.E.; Robb, M.A.; Cheeseman, J.R.; Scalmani, G.; Barone, V.; Petersson, G.A.; Nakatsuji, H.; et al. *Gaussian 16, Revision C.01*; Gaussian, Inc.: Wallingford, CT, USA, 2016.
50. Becke, A.D. Density-functional thermochemistry. III. The role of exact exchange. *J. Chem. Phys.* **1993**, *98*, 5648–5652. [CrossRef]
51. Lee, C.T.; Yang, W.T.; Parr, R.G. Development of the Colle-Salvetti correlation-energy formula into a functional of electron density. *Phys. Rev. B* **1988**, *37*, 785–789. [CrossRef]
52. Martin, R.L. Natural Transition Orbitals. *J. Chem. Phys.* **2003**, *118*, 4775–4777. [CrossRef]
53. Le Bahers, T.; Adamo, C.; Ciofoni, I. A Qualitative Index of Spatial Extent in Charge-Transfer Excitations. *J. Chem. Theory Comput.* **2011**, *7*, 2498–2506. [CrossRef] [PubMed]
54. Dennington, R.; Keith, T.A.; Millam, J. *GaussView, Version 6.0*; Semichem Inc.: Shawnee Mission, KS, USA, 2016.
55. Chemcraft—Graphical Software for Visualization of Quantum Chemistry Computations. Available online: https://www.chemcraftprog.com (accessed on 5 May 2021).
56. Lu, T.; Chen, F. Multiwfn: A Multifunctional Wavefunction Analyzer. *J. Comput. Chem.* **2012**, *33*, 580–592. [CrossRef] [PubMed]
57. Kouki, N.; Trabelsi, S.; Seydou, M.; Maurel, F.; Tangour, B. Internal path investigation of the acting electrons during the photocatalysis of panchromatic ruthenium dyes in dye-sensitized solar cells. *Comptes Rendus Chim.* **2019**, *22*, 34–45. [CrossRef]
58. Qu, Z.-W.; Kroes, G.-J. Theoretical Study of the Electronic Structure and Stability of Titanium Dioxide Clusters $(TiO_2)_n$ with n =1–9. *J. Phys. Chem. B* **2006**, *110*, 8998–9007. [CrossRef] [PubMed]
59. De Angelis, F.; Fantacci, S.; Selloni, A. A Time-dependent density functional theory study of the absorption spectrum of [Ru(4,40-COOH-2,20-bpy)2(NCS)2]in water solution: Influence of the pH. *Chem. Phys. Lett.* **2004**, *389*, 204–208. [CrossRef]
60. Albright, T.A.; Burdett, J.K.; Whangobo, M.-H. *Orbital Interactions in Chemistry*, 2nd ed.; John Wiley & Sons: Hoboken, NJ, USA, 2013; pp. 416–422.
61. Sundari, C.D.D.; Martoprawiro, M.A.; Ivansyah, L. A DFT and TDDFT Study of PCM Effect on N3 Dye Absorption in Ethanol Solution. *J. Phys. Conf. Ser.* **2017**, *812*, 12068. [CrossRef]
62. Neugebauer, J.; Reiher, M.; Kind, C.; Hess, B.A. Quantum Chemical Calculation of Vibrational Spectra of Large Molecules—Raman and IR Spectra for Buckminsterfullerene. *J. Comp. Chem.* **2002**, *23*, 895. [CrossRef]
63. Lombardi, J.R.; Birke, R.L. A Unified Approach to Surface-Enhanced Raman Spectroscopy. *J. Phys. Chem. C* **2008**, *112*, 5605. [CrossRef]
64. Albrecht, A.C. On the Theory of Raman Intensities. *J. Chem. Phys.* **1961**, *34*, 1476. [CrossRef]
65. Birke, R.L.; Lombardi, J.R. Relative Contributions of Franck-Condon to Herzberg-Teller terms in charge transfer surface-enhanced Raman scattering spectroscopy. *J. Chem. Phys.* **2020**, *152*, 224107. [CrossRef] [PubMed]
66. Chen, X.-K.; Coropceanu, V.; Brédas, J.-L. Assessing the nature of the charge-transfer electronic states in organic solar cells. *Nat. Commun.* **2018**, *9*, 5295. [CrossRef]

nanomaterials

Article

SERS-Based Aptasensor for Rapid Quantitative Detection of SARS-CoV-2

Elena Zavyalova [1,*], Oganes Ambartsumyan [2], Gleb Zhdanov [1], Dmitry Gribanyov [3], Vladimir Gushchin [4], Artem Tkachuk [4], Elena Rudakova [5], Maria Nikiforova [4], Nadezhda Kuznetsova [4], Liubov Popova [4], Bakhtiyar Verdiev [4], Artem Alatyrev [4], Elena Burtseva [4], Anna Ignatieva [4], Anna Iliukhina [4], Inna Dolzhikova [4], Alexander Arutyunyan [6], Alexandra Gambaryan [7] and Vladimir Kukushkin [3,*]

[1] Chemistry Department, Lomonosov Moscow State University, 119991 Moscow, Russia; gleb.zhdanov@chemistry.msu.ru
[2] Department of Microbiology, Virology and Immunology, I.M. Sechenov First Moscow State Medical University, 125009 Moscow, Russia; ogan-mail@mail.ru
[3] Institute of Solid State Physics of Russian Academy of Science, 142432 Chernogolovka, Russia; digrib@gmail.com
[4] National Research Center for Epidemiology and Microbiology Named after the Honorary Academician N. F. Gamaleya, 123098 Moscow, Russia; wowaniada@gmail.com (V.G.); artem.p.tkachuk@gmail.com (A.T.); marianikiforova@inbox.ru (M.N.); nadyakuznetsova0@gmail.com (N.K.); ljubovprokudina@gmail.com (L.P.); yuryevpolskei@yandex.ru (B.V.); artem.alatyrev@gmail.com (A.A.); elena-burtseva@yandex.ru (E.B.); valgella@yandex.ru (A.I.); sovanya97@yandex.ru (A.I.); i.dolzhikova@gmail.com (I.D.)
[5] Institute of Physiologically Active Compounds of Russian Academy of Science, 142432 Chernogolovka, Russia; evladru@mail.ru
[6] Belozersky Research Institute of Physico-Chemical Biology, Lomonosov Moscow State University, 119991 Moscow, Russia; Alarut@genebee.msu.ru
[7] Chumakov Federal Scientific Center for Research and Development of Immune and Biological Products RAS, 108819 Moscow, Russia; al.gambaryan@gmail.com
* Correspondence: zlenka2006@gmail.com (E.Z.); kukushvi@mail.ru (V.K.)

Citation: Zavyalova, E.; Ambartsumyan, O.; Zhdanov, G.; Gribanyov, D.; Gushchin, V.; Tkachuk, A.; Rudakova, E.; Nikiforova, M.; Kuznetsova, N.; Popova, L.; et al. SERS-Based Aptasensor for Rapid Quantitative Detection of SARS-CoV-2. *Nanomaterials* **2021**, *11*, 1394. https://doi.org/10.3390/nano11061394

Academic Editors: Ronald Birke and Bing Zhao

Received: 30 April 2021
Accepted: 22 May 2021
Published: 25 May 2021

Abstract: During the COVID-19 pandemic, the development of sensitive and rapid techniques for detection of viruses have become vital. Surface-enhanced Raman scattering (SERS) is an appropriate tool for new techniques due to its high sensitivity. SERS materials modified with short-structured oligonucleotides (DNA aptamers) provide specificity for SERS biosensors. Existing SERS-based aptasensors for rapid virus detection are either inapplicable for quantitative determination or have sophisticated and expensive construction and implementation. In this paper, we provide a SERS-aptasensor based on colloidal solutions which combines rapidity and specificity in quantitative determination of SARS-CoV-2 virus, discriminating it from the other respiratory viruses.

Keywords: SARS-CoV-2; COVID-19; SERS; silver colloids; aptamer; optical sensor; respiratory viral infections

1. Introduction

In late 2019 in Wuhan (Hubei, China), a new strain of coronavirus SARS-CoV-2 was first detected, which caused a pandemic of acute respiratory disease, named coronavirus disease 2019 (COVID-19). The virus is rapidly spread and has already claimed the lives of more than two million people all over the world. It also made a significant impact on global economy, put pressure on healthcare, and affected social life. The disease is characterized by progressive inflammatory pulmonary cellular infiltration, hypoxemia, and the development of an acute respiratory distress syndrome (ARDS), which could lead to death.

The situation is complicated by the fact that many infected patients have very mild symptoms or are completely asymptomatic during the entire period of infection, but can still transmit the disease to others [1,2]. Additionally, symptoms of COVID-19 are similar to other respiratory illnesses, so it needs to be differentiated from bacterial and other viral pneumonia caused by influenza virus, adenovirus, parainfluenza, or respiratory syncytial virus (RSV) for clinical diagnostic [3].

Traditional detection methods such as quantitative polymerase chain reaction with a reverse transcription (RT-qPCR) and enzyme-linked immunosorbent assay (ELISA) require qualified personnel, laboratory with separate areas, often expensive reagents, and complicated sample preparation (especially for the PCR), in which errors can lead to false positive or false negative results [4,5]. Lateral flow assay (LFA) tests are also widespread due to the COVID-19 pandemic. Their advantages include speed of analysis, as well as no need for qualified personnel and a special room; therefore, a person can test himself for disease. However, these tests can give false negative results more often as compared to the above methods due to their lower sensitivity [6–8].

Due to rapid spread of COVID-19 around the world and similar symptoms to other viral and bacterial pneumonia, an interest in the development of fast, easy-to-use, and selective methods for biosensor diagnostics of respiratory diseases has increased.

Biosensors present a great interest for scientists and clinicians in diagnostics due to their advantages over the existing laboratory methods such as RT-qPCR and antibody-based techniques (enzymatic, fluorescent, and chemiluminescent). They have similar sensitivity but require less time and no advanced laboratory equipment with qualified personnel. They are also as fast but more sensitive than LFAs [9].

Among the promising devices, there are biosensors based on surface-enhanced Raman scattering (SERS). They can be classified into label-free and reporter-based platforms. The using of label-free schemes is limited because, usually, biological object is analyzed in a mixture of other biological substances, which causes overlapping of spectral lines and difficulty in identifying substances [10]. More promising is using commercially available Raman-active dyes, which are widely used as reporter molecules [11,12]. Another important component of the biosensor is recognizing elements such as antibodies or aptamers, which allows capture of certain objects from a biological mixture.

This capture, in turn, results in a change of a certain physical property, not limited to a change in Raman spectra [13]. Moreover, the introduction of biosensors on the surface of plasmonic nanoparticles often leads to optical detectors that, upon binding, make the target change itscolour [14]. In recent years, there has been a growing interest in using aptamers as biorecognizing elements. In particular, aptamer-based biosensors provide a great level of specificity and variability of modifications. Aptamers are an alternative to antibodies for their comparable qualities in detection [15].

For the existing SERS-based aptasensors for detection of respiratory viruses, there are several published approaches. Negrie et al. suggested an application of a polyvalent DNA aptamer to influenza nucleoprotein. Sigmoidal dependence of direct SERS intensity was a function of aptamer concentration on Ag substrate from 1 to 5000 nM. The binding of the target and aptamer changed a secondary structure, which was sensed by SERS. Incubation with a viral content took more than 8 h for registering SERS spectra of the aptamer, which limits practical implementation [16]. In the latest works devoted to influenza detection, Chen et al. developed an impressive aptasensor for influenza A (H1N1) with LOD of 97 PFU/mL and 20 min per an assay with a reverse monotonous dependence on a viral load. However, preparation of an advanced 3D nano-popcorn SERS substrate demanded more than 7 h [17].

For the newest aptasensors for SARS-CoV-2, Stanborough et al. compared SPR, BLI, and SERS in performance of direct spike protein detection. SERS-detection showed the lowest LOD of 1 fM and dynamic range of six orders. It required more than 2 h of sample preparation and did not imply whole viral particles detection, which has less clinical significance [18]. Yang et al. [19] presented a human angiotensin-converting-enzyme-2functionalized gold "virus traps" nanostructure as a very sensitive SERS biosensor. This optical sensor was used for selective capture and rapid detection of S-protein (expressed SARS-CoV-2) in the contaminated water.

In our previous work, we compared the use of solid-state SERS substrates and colloidal SERS particles to detect influenza A virus. On solid-state SERS substrates, we managed to achieve LOD for influenza A virus of 10^4 VP/mL with less than 15 min for an assay. We applied SERS substrate with immobilized primary DNA aptamer RHA0385 with broad specificity towards to influenza A virus hemagglutinin on silver and secondary labeled aptamer for "sandwich" assembly. This approach allowed to detect whole viral particles with hemagglutination activity. However, the dependence was not monotonous, which allowed only qualitative interpretation of the result [20].

In our second work of influenza detection, we switched from the solid SERS substrate to the colloidal one with the same "sandwich" approach and aptamers. This allowed us to create a simple and rapid "one-pot" technique with a monotonous dependence but with a modest dynamic range of $2 \times 10^5 - 2 \times 10^6$ VP/mL [21].

For this work, we used a colloidal SERS-based aptasensor with a new setup including changes in nanoparticle aggregation, aptamer, and target virus, i.e., SARS-CoV-2 virus. The aptamer was RBD-1C, which was shown to have high affinity to receptor binding domain (RBD) of surface S-protein of SARS-CoV-2 [22].

2. Materials and Methods

2.1. Procedure of AgNPs Synthesis

Preparation of the silver colloids by reducing a silver nitrate solution with hydroxylamine hydrochloride was conducted in concordance with the method [23]. Silver nitrate ($AgNO_3$, CAS 7761-88-8), hydroxylamine hydrochloride (NH_2OH-HCl, CAS 5470-11-1) were of the highest purity available; sodium hydroxide (NaOH, CAS 1310-73-2), sodium chloride (NaCl, CAS 7647-14-5), xanthine (CAS 69-89-6)were analytical-grade and used without further purification. All the products were purchased from Aldrich (Sigma-Aldrich Inc., St. Louis, MO, USA). The stock solutions of reagents and silver sols were prepared with water from a Milli-Q system (18.2 MΩ·cm resistivity at 25 °C).

The AgNPs were prepared at room temperature under vigorous stirring by rapid addition of a small volume of a concentrated silver solution (10 mL) to a large volume of a less-concentrated hydroxylamine hydrochloride/sodium hydroxide solution (90 mL). The concentration of $AgNO_3$ and NH_2OH-HCl/NaOH was 10^{-3} M and $1.5 \times 10^{-3}/3 \times 10^{-3}$ M, respectively, in the final reaction mixture. It was kept under stirring for 30 min once the addition of silver nitrate was complete. A yellow-brown colored transparent colloidal suspension with no sediment was obtained. The synthesized colloids were stable for several weeks under refrigeration conditions at $+6 - +8$ °C.

UV-Vis spectra of AgNPs were recorded on a spectrometer Genesis 10S UV-Vis (Thermo-Fisher Scientific, Madison, WI, USA). The morphology of silver nanoparticles was studied by SEM (Scanning Electron Microscopy), performed using a Supra 50VP electron microscope (Zeiss, Germany) with a resolution of 1.5 nm. The mean size and ζ-potential of AgNP were determined by DLS (Dynamic Light Scattering), performed using Zetasizer Nano ZS (Malvern, Worcestershire, UK).

2.2. Viruses

Vero E6 (ATCC CRL-1586) cell line was obtained from the Chumakov Federal Scientific Center for Research and Development of Immune and Biological Products RAS, Russia, and was maintained in complete Dulbecco's modified Eagle's medium (DMEM), containing 10% fetal bovine serum (FBS, HyClone, Logan, Utah, USA), L-glutamine (4 mM), and penicillin/streptomycin solution (100 IU/mL; 100 µg/mL) (PanEco, Moscow, Russia). SARS-CoV-2 isolate PMVL-5 (GISAID accession EPI_ISL_470899) was isolated in May 2020 from a nasopharyngeal swab specimen taken from a 22-year-old female. The nasopharyngeal swab was inoculated on Vero E6 (non-human primate kidney) cells. The inoculated cells were monitored for cytopathic effects by light microscopy and cytopathic effects were detected at 72 h post inoculation. Virus was passaged three times before the experiments to pile the virus. Viral titer of 4.6×10^6 TCID50/mL was determined as

TCID50 by endpoint dilution assay. Virus titer was calculated using the Reed and Muench method [24]. All experiments with the live SARS-CoV-2 followed the approved standard operating procedures of the NRCEM biosafety level 3 facility.

Influenza viruses and Newcastle disease virus were provided by the Chumakov Federal Scientific Center for Research and Development of Immune and Biological Products of the Russian Academy of Sciences. The following strains were studied: influenza A virus (IvA) A/FPV/Rostock/34 R6p (256 HAU/mL in stock solution); influenza B virus (IvB) B/Victoria/2/1987 (2000 HAU/mL in stock solution); Newcastle disease virus(NDV) (256 HAU/mL in stock solution). Virus stocks were propagated in the allantoic cavity of 10-day-old embryonated specificpathogen-free chicken eggs. Eggs were incubated at 37 °C, cooled at 4 °C 48 h post-infection, and harvested 16 h later. The study design was approved by the Ethics Committee ofthe Chumakov Institute of Poliomyelitis and Viral Encephalitides, Moscow, Russia (Approval #4 from 2 December 2014). Viruses were inactivated via the addition of 0.05% (*v*/*v*) glutaric aldehyde, preserved via the addition of 0.03% (*w*/*v*) NaN_3, and stored at +4 °C.

Adenovirus type 6 (AdV) Strain Tonsil 99 (Bialexa, Russia) and respiratory syncytial virus (RSV) (Bialexa, Russia) were inactivated by treatment with Thimerosal and beta propiolactone. Viral content of RSV was 1 mg/mL ($2 \cdot 10^{12}$ VP/mL); viral content of AdV was 1.9 mg/mL ($4 \cdot 10^{12}$ VP/mL). Viral particle concentrations (VP/mL) were calculated from the protein concentration.

2.3. Aptamers and Their Assembly

The following oligonucleotides were studied. Biotin-RBD-1C (Biotin-5′-CAGCAC CGACCTTGTGCTTTGGGAGTGCTGGTCCAAGGGCGTTAATGGACA-3′ from Synthol, Moscow, Russia), Biotin-RBD-1C-sh (Biotin-5′-TTTGGGAGTGCTGGTCCAAGGGCGTTAA TGGACA-3′ from Synthol, Moscow, Russia), BDP FL-RBD-1C (Bodipy FL-5′-CAGCACCGA CCTTGTGCTTTGGGAGTGCTGGTCCAAGGGCGTTAATGGACA-3′ from Lumiprobe, Moscow, Russia). To assemble the structure of the aptamers, the following algorithm was used. Biotinylated aptamers were prepared in 2 μM concentrations in the buffer containing 10 mM tris-HCl pH 7.0, 140 mM NaCl, and 10 mMKCl. Bodipy FL labeled aptamer was prepared in 2 μM concentrations in the buffer containing 10 mM PBS. The solutions were heated at 95 °C for 5 min and cooled at room temperature.

2.4. Circular Dichroism and UV-Spectroscopy

A 2 μM aptamer solution was placed in a quartz cuvette with 1 cm path. Circular dichroism (CD) spectra were acquired using CD spectrometer Chirascan (Applied Photophysics, Leatherhead, UK) and a dichrograph MARK-5 (Jobin-Yvon; Montpellier, France) equipped with a thermoelectric temperature regulator. The spectra were acquired in the range of 220–360 nm.

2.5. Determination of Aptamer Affinity to S-Protein and Binding to SARS-CoV-2 Virus

Kinetic constants of association and dissociation of the RBD-S-protein complexes were determined using interferometer Blitz (Forte-Bio, Fremont, CA, USA). Streptavidin biosensors from Forte-Bio (USA) were used. The biosensors were hydrated for 10 min in the buffer. The biotinylated aptamer was loaded on the biosensor from 1 μM solution for 120 s. The binding experiments were performed as following:

(1) baseline in the buffer during 30 s;
(2) association stage in 80–1200nM RBD of S-protein (from HyTest, Turku, Finland) solution during 200 s;
(3) dissociation stage in the buffer during 300 s.

Kinetic constants were calculated from exponential approximations of the curves [25]. Several experiments were performed to test the affinity of the aptamers to inactivated SARS-CoV-2 virus, including experiments on Photonic Crystal Surface Mode (PC SM)-

based biosensor "EVA 2.0" (Institute of Spectroscopy, Russian Academy of Sciences, Troitsk, Moscow, Russia).

The experiments on "EVA 2.0" with viruses were performed as following:multilayered glass SiO_2 I Ta_2O_5 I SiO_2 was silanized with APTES from the one side by hydrolisation. The other side of the glass was treated with oil to the panel of the device with a red laser. A glass cell with hosepipes was installed on top of the glass. A disc pump was connected to the outlet hose. At the beginning of the experiment, bidistilled water was run through the entire system to avoid the bubbles' formation and to elevate of measurement accuracy. The ideal interference pattern obtained could be seen through the video camera set up on the device.

The silanized side of the glass was treated with the following solutions at room temperature:

(1) An amount of 0.1 M NaH_2PO_4 pH = 6.2 for 2 min.
(2) The activation of -NH_2 groups of APTES by EDC (50 mg/mL) and NHS (50 mg/mL) for 20 min.
(3) An amount of 50 mM MES pH~5.0 for 5 min.
(4) A covalent conjugation of streptavidin to the activated -NH_2 groups. Streptavidin (40 μg/mL) in 50 mM MES pH~5.0 for 30 min.
(5) Washing with 50 mM MES pH~5.0 for 2 min.
(6) Washing with 5 mM glycine in 25 mM HCl (pH~2) for 1–2 min.
(7) Washing with 10 mM Tris-HCl (pH = 7.4), 140 mM NaCl, 10 mM KCl for 5 min.
(8) Loading of biotinylated aptamer to SARS-CoV-2 (RBD-1C) for 15–20 min.
(9) Washing with 10 mM Tris-HCl (pH = 7.4), 140 mM NaCl, 10 mM KCl for 5 min.
(9a) Interaction with cell medium for 3 min.
(9b) Interaction with SARS-CoV-2 virus with a titer of $0.22 \cdot 10^6$ TCID50/mL or $0.11 \cdot 10^6$ TCID50/mL for 3 min.
(10) Dissociation in 10 mM Tris-HCl (pH = 7.4), 140 mM NaCl, 10 mM KCl for 5 min.

2.6. SERS Measurements

The SERS spectra were acquired using a handheld Raman analyzer RaPort (Enhanced Spectrometry, Inc., San Jose, CA, USA) with a laser wavelength of 532 nm and a working output power of 38 mW. The spectrometer had a spectral resolution of 4–6 cm^{-1} and a spectral range of 160–4000 cm^{-1}. The recording time of a single spectrum was 400 ms with 20 averages. The measurements were carried out in glass vials with a volume of 1.5 mL (Akvilon, Moscow, Russia). The focus of the laser beam coincided with the center of the vial.

We took 150; 50; 25; 12.5; 9; 7.5; 6; 4.5; 3; 1.5; 0.75; 0.37; and 0.15 μl of a cell culture medium with SARS-CoV-2, control viruses (RSV, IvA, IvB, NDV, AdV), or a virus-free cell culture medium. The cell culture medium was the same for all viruses for a more objective comparison. The control viruses were prepared in a concentration of 10^8 VP/mL, supposing that the ratio between TCID50/mL and VP/mL is about 1 to 100 as was estimated for fresh influenza viruses by Kramberg et al. [26]. Then, we added 4 μL of 2 μM BDP FL-RBD-1C. For the volumes of 25 μL and less, we added PBS to obtain a total volume of 50 μL before the addition of the labeled aptamer. After 5 min of the incubation, we added 250 μL of PBS and injected the solution to 196 μL of AgNPs (the total volume was 500 μL). SERS spectra was registered at 1 min after the aggregation step.

3. Results

3.1. Characterization of Silver Colloidal Nanoparticles

In this work, we used hydroxylamine-reduced AgNPs [23] as SERS substrate. These silver colloids are stable and highly SERS-active with good reproducibility in the obtained enhancement factors, as well as having the possibility of long-term (about a month) storage without loss of efficiency.

To characterize the morphology of the produced colloids, UV-Vis spectroscopy (Figure S1) and DLS (Figure S2) were used. The absorption maximum of the measured UV-Vis spectrum of the colloidal solution provides information on the average particle size, whereas its full width at half-maximum (FWHM) can be used to estimate particle dispersion. The absorption maximum was found at 394 nm with a FWHM of approximately 55 nm. The mean size according to the number distribution of DLS is 4.8 nm, whereas the intensity distribution gives the size of 44 nm, being in agreement with the value calculated from spectral characteristics of NP. These results have been confirmed by SEM (Figure S3). The AgNPs are supposed to have no interaction with the aptamer as both have strong negative charge. ζ-potential of NPs is −54 + −18 mV (Figure S4).

Xanthine, purine base, was selected as a test compound to investigate the effectiveness of silver colloids in enhancing SERS spectra. The spectra of the test substance xanthine at different ionic strengths of the solution are shown in Figure S5.

3.2. Structural Characterization of Aptamer RBD-1C

The structure of the aptamer RBD-1C was supposed by Song et al. [22]; it consists of two hairpins that are end-to-end stacked (Figure 1A). The RNA fold webserver [27] provided the same folding for the sequence of RBD-1C. DNA duplexes are known to have hyperchromic effect during unfolding in UV-spectra at the wavelength of 260 nm [28]. However, no notable changes in UV-spectra were tracked during RBD-1C melting in potassium buffer (Figure S6). Circular dichroism (CD) spectra were recorded to identify the other possible topologies of RBD-1C (Figure 1B).

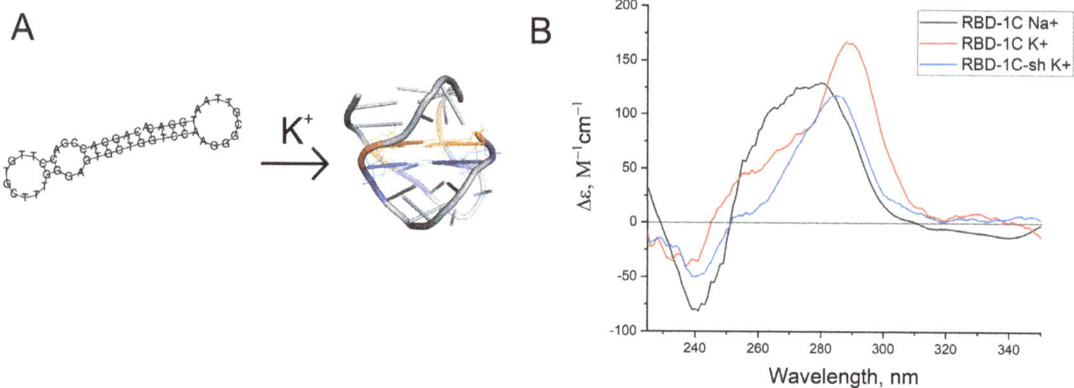

Figure 1. (**A**) Putative structures of the aptamer RBD-1C. (**B**) Circular dichroism spectra of RBD-1C in 140 mM NaCl ('Na^+'), RBD-1C, and RBD-1C-sh in 140 mM NaCl and 10 mM KCl ('K^+').

The spectra contained a positive peak at 290 nm, a shoulder at 254 nm, and a slight negative peak at 230 nm. These CD spectra could not be attributed to duplexes [29]; similar spectra were reported for two-quartet antiparallel G-quadruplex with a diagonal loop (Figure 1A) [30]. CD spectra in sodium buffer had a shoulder at 290 nm and major peak at 270–280 nm that correspond to a preferable DNA duplex. Thus, an equilibrium between two conformations was observed. A shortened version of the aptamer was studied to reveal whether G-quadruplex is a recognizing element in this aptamer. Seventeen nucleotides from the 5′-end of the aptamer were excluded (oligonucleotide RBD-1C-sh). This oligonucleotide had CD spectra that are close to RBD-1C. Both oligonucleotides were studied in affinity experiments.

3.3. Affinity of RBD-1C and Its Shortened Version to RBD of S-Protein and the Whole Virus

The affinity of RBD-1C to the recombinant protein has been previously estimated by Song et al. [22]. Here, we estimated it once more and compared the affinities of RBD-1C with its truncated variant. The sensorgrams are shown in Figure 2.

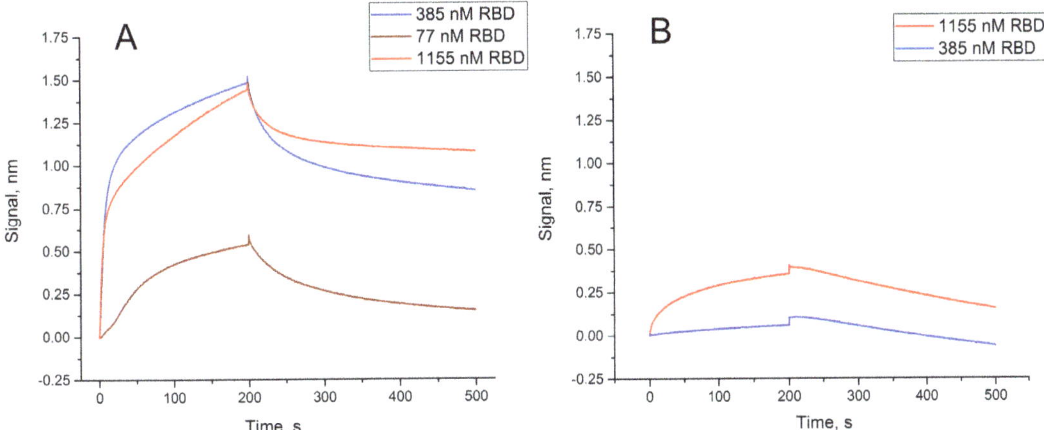

Figure 2. Sensorgrams of the interaction between soluble RBD of S-protein and immobilized oligonucleotides: (**A**) RBD-1C; (**B**) RBD-1C-sh. The concentrations of protein are provided in the graphs.

Dissociation constant of RBD-1C was 13 ± 2 nM (Table 1),which agrees well with the previous value of 5.8 ± 0.8 nM [19]. The truncated variant, RBD-1C-sh, had a dissociation constant of 330 ± 60 nM that corresponds to low specific binding.

Table 1. The parameters of the complexes of RBD of S-protein with aptamer RBD-1C and its shortened version, RBD-1C-sh. Kinetic constants of association (k_a) and dissociation (k_d) as well as equilibrium dissociation constants (K_d) are provided.

Aptamer	k_a, $M^{-1}s^{-1}$	k_d, s^{-1}	K_d, M
RBD-1C	$(1.7 \pm 0.8)\cdot 10^5$	$(2.17 \pm 0.02)\cdot 10^{-3}$	$(1.3 \pm 0.2)\cdot 10^{-8}$
RBD-1C-sh	$(1.7 \pm 0.2)\cdot 10^4$	$(5.6 \pm 0.4)\cdot 10^{-3}$	$(3.3 \pm 0.6)\cdot 10^{-7}$

Thus, the unique spatial structure of RBD-1C is crucial for high affine recognition of the RBD of S-protein. Next, the ability of binding the whole virus was tested for both RBD-1C and RBD-1C-sh. RBD-1C showed excellent curves for both concentrations of the virus; whereas RBD-1C-sh provided low-efficient binding of the viruses (Figure 3). It could be concluded that the active aptamer structure is a DNA hairpin of RBD-1C, whereas G-quadruplexes are low-active additives. The content of the virus was estimated in TCID50 that corresponds to virulent particles.

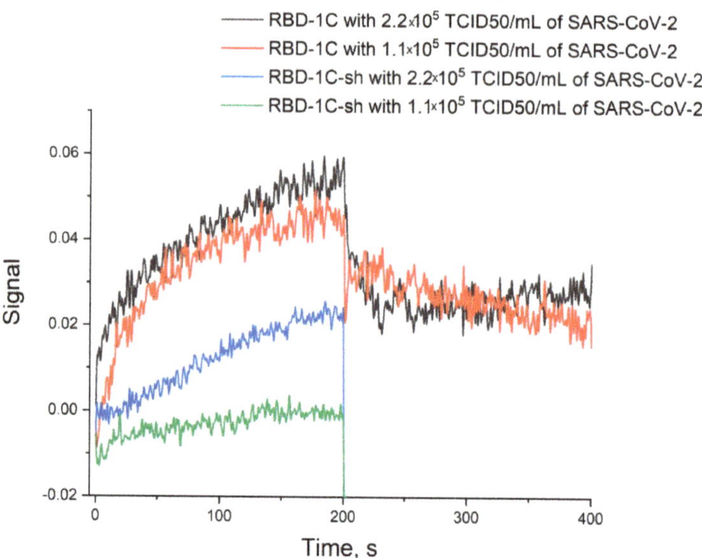

Figure 3. Sensorgrams of the interaction between SARS-CoV-2 viruses and immobilized oligonucleotides RBD-1C and RBD-1C-sh. The concentrations of the virus are provided.

3.4. Specificity of SERS Aptasensor

For each virus studied, the concentration dependence of the SERS signal was investigated. The cell culture medium was used as a control. In the observed SERS spectra in the experimental and control samples, the intensity of the main Raman peak of the Bodipy FL label 587 cm^{-1} was studied (Figure 4A). As a result, the concentration dependence of the intensity ratio of the SERS line (587 cm^{-1}) in the experimental and control samples was constructed (Figure 4C,D).

Figure 4C demonstrates the performance of the system for detecting the inactivated SARS-CoV-2. The specificity was tested for RSV, IvA, IvB, NDV, and AdV. For qualitative comparison, control viruses were reduced to uniform concentrations and concentration curves were constructed for them. Calibration curve for SARS-CoV-2 is ranging from 5.5×10^4 to 1.4×10^6 TCID50/mL.

Our hypothesis explaining the behavior of the curves for SARS-CoV-2 and for non-specific viruses (Figure 4C) and describing the mechanism of SERS signal generation is as follows:

- It is known [31,32] that at high ionic strength, silver nanoparticles interact well with the surface proteins of the virus, forming aggregates on the virus particles. With an increase in the concentration of proteins, the aggregates of silver nanoparticles create inhomogeneities of the electromagnetic field with a locally high density («hot spots») near their surface.
- Due to significant differences in the number of the reporter molecules near non-specific viral particles and SARS-CoV-2, the SERS signal decreases with increasing concentration of the non-specific virus, and in the case of SARS-CoV-2, it increases.

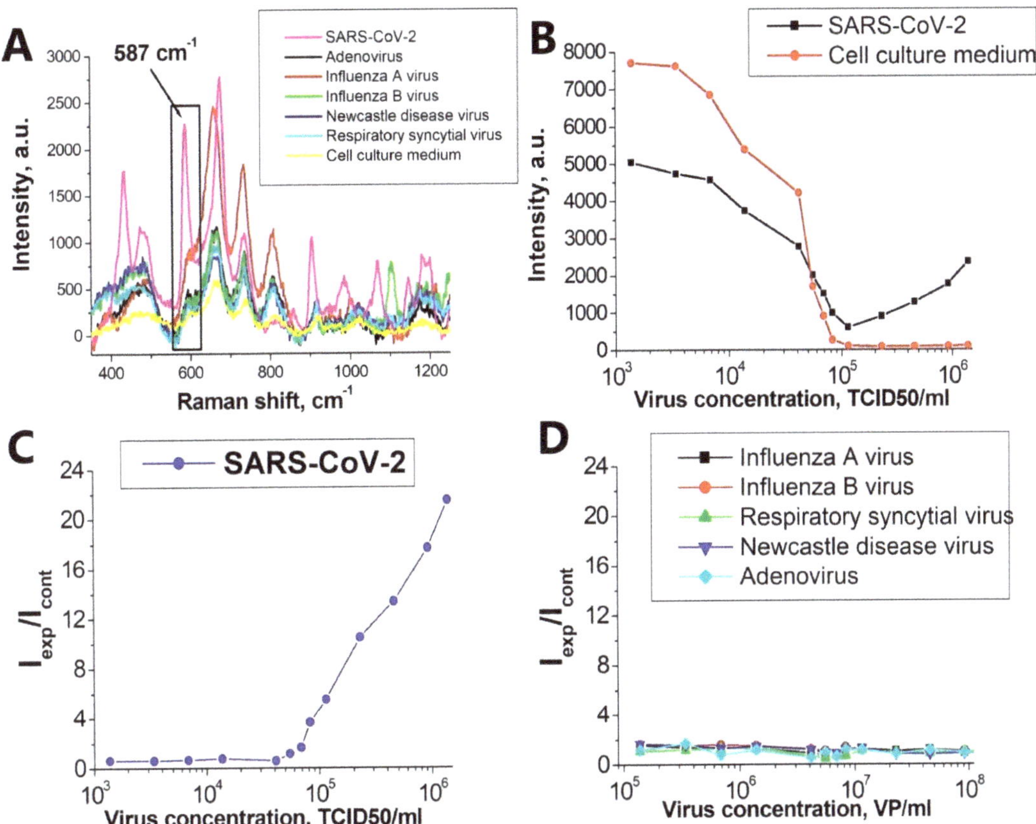

Figure 4. Detection of SARS-CoV-2. (**A**) SERS spectra of an experimental virus sample (SARS-CoV-2) at a concentration of 4.6×10^5 TCID50/mL; virus-free cell culture medium in the same dilutions and control viruses in the same concentrations. (**B**) The dependence of the peak intensity of 587 cm^{-1} on the concentration of SARS-CoV-2 and for the virus-free cell culture medium in the same dilutions. (**C**) Concentration curve for SARS-CoV-2. (**D**) Concentration curve for control viruses.

Figure 5 shows the scheme of the experiment and explains the principle of operation of the SERS-aptasensor. The target virus (SARS-CoV-2 virion, left side) accumulates more labeled aptamers due to the specific interaction mentioned earlier, while control viruses (i.e., influenza A virion, right side) have fewer labeled aptamers for the same enhancement field from AgNPs aggregates. This resultsin an increase of SERS intensity with higher concentrations of SARS-CoV-2 and in a decrease for control viruses.

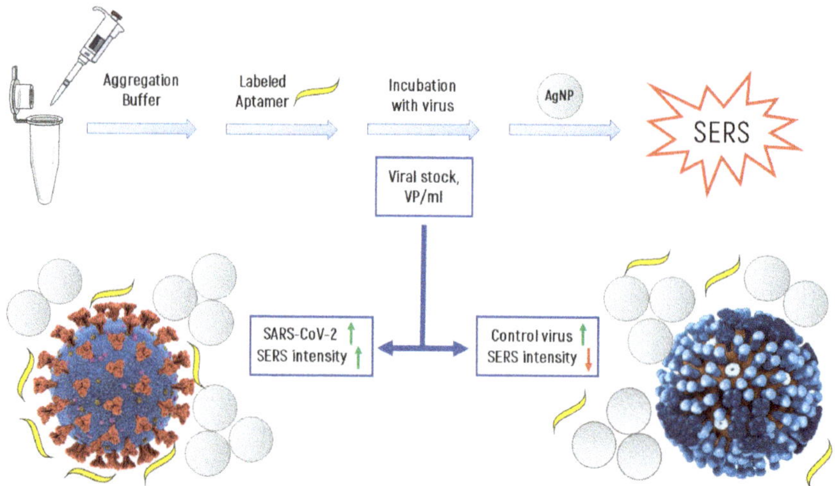

Figure 5. The scheme of the experiment and the formation of the SERS signal.

4. Discussion

SARS-CoV-2 virus has many common symptoms inherent in other respiratory viral infections, therefore, it is often identified only in the late stages of the disease. In this regard, the creation of a simple and cheap method and device for its rapid detection may slow down such a rapid increase in the spread of the pandemic.

There are many different types of SERS substrates [33–36] and schemes for creating aptasensors based on them [20,37–40]. However, to create a quantitative method for the selective detection of SARS-CoV-2 from a group of other respiratory viral infections, this method should be the cheapest, simplest, and most reproducible. The reproducibility of SERS sensors can be achieved only in the case of using either nanoperiodic SERS structures obtained using electron lithography methods [41,42] (however, in this case, the cost of their production is high and therefore the method cannot compete with ELISA), or colloidal SERS solutions [23,43]. Colloidal solutions are easy to synthesize, homogeneous, and allow to work on a handheld Raman spectrometer with high laser radiation powers for a long exposure time.

In this paper, we have developed a very simple rapid method for determining SARS-CoV-2, based on the scheme of a direct method for assembling SERS aptasensor using colloidal SERS. The method presented in this article is simple («one-step»), fast (7 min), and has high sensitivity (LOD of 5.5×10^4 TCID50/mL) and specificity.

Test specificity is a vital problem in diagnostics and the following treatment. In all existing detection methods, there is a trade-off between sensitivity and specificity. In the ideal situation, both diagnostic sensitivity and diagnostic specificity are 100%. However, the fact is that this situation is rarely reached and intention to increase the diagnostic sensitivity results in a decrease in diagnostic specificity. For tests in which false positives are unacceptable but false negatives may occur, high diagnostic specificity is required. It should be noticed that the more steps, materials, and hard-managed equipment a new method requires, the greater the degree of cross-reactivity in the approach.

Therefore, it is extremely important to develop new methods for selective and rapid detection of other viral respiratory diseases, such as: influenza virus, adenovirus, respiratory syncytial virus, and Newcastle disease virus. The authors of this article are already working on the creation of optical aptasensors for the selective detection of these viruses.

Supplementary Materials: The following are available online at https://www.mdpi.com/article/10.3390/nano11061394/s1, Figure S1: UV-Vis spectroscopy of colloidal silver solutions, Figure S2: Dynamic Light Scattering (DLS) of colloidal silver solutions, Figure S3: SEM of AgNPs, Figure S4: ζ-potential of AgNPs, Figure S5: Spectra of the test substance (2 mkM xanthine) at different ionic strengths of the solution, Figure S6: UV-spectra of RBD-1C (during melting in potassium buffer).

Author Contributions: Conceptualization, E.Z., O.A., V.G., A.T., V.K.; methodology, E.Z., O.A., D.G., E.R., M.N., N.K., L.P., A.I. (Anna Ignatieva), I.D., A.A. (Alexander Arutyunyan); formal analysis, G.Z., B.V., A.A. (Artem Alatyrev), A.I. (Anna Iliukhina), A.G.; writing—original draft preparation, E.Z., O.A., G.Z., V.K.; writing—review and editing, E.Z., G.Z., D.G., A.T., B.V., E.B., V.K.; project administration, E.Z., V.K.; funding acquisition, V.K. All authors have read and agreed to the published version of the manuscript.

Funding: Work on the creation of highly specific detection of SARS-CoV-2 was supported by the Russian Foundation for Basic Research (project № 20-04-60077). The authors are grateful to the Russian Science Foundation (grant number 19-72-30003) for the development of effective SERS substrates.

Institutional Review Board Statement: The study was conducted according to the guidelines of the Declaration of Helsinki, and approved by the Ethics Committee of the Chumakov Institute of Poliomyelitis and Viral Encephalitides, Moscow, Russia (Approval #4 from 2 December 2014).

Informed Consent Statement: Not applicable.

Conflicts of Interest: The authors declare no conflict of interest. The funders had no role in the design of the study; in the collection, analyses, or interpretation of data; in the writing of the manuscript; or in the decision to publish the results.

References

1. Johansson, M.A.; Quandelacy, T.M.; Kada, S.; Prasad, P.V.; Steele, M.; Brooks, J.T.; Slayton, R.B.; Biggerstaff, M.; Butler, J.C. SARS-CoV-2 Transmission from People Without COVID-19 Symptoms. *JAMA Netw. Open.* **2021**, *4*, e2035057. [CrossRef] [PubMed]
2. Ye, F.; Xu, S.; Rong, Z.; Xu, R.; Liu, X.; Deng, P.; Liu, H.; Xu, X. Delivery of infection from asymptomatic carriers of COVID-19 in a familial cluster. *Int. J. Infect. Dis.* **2020**, *94*, 133–138. [CrossRef]
3. Sharma, R.; Agarwal, M.; Gupta, M.; Somendra, S.; Saxena, S. Clinical Characteristics and Differential Clinical Diagnosis of Novel Coronavirus Disease 2019 (COVID-19). Coronavirus Disease 2019 (COVID-19). In *Medical Virology: From Pathogenesis to Disease Control*; Saxena, S., Ed.; Springer: Singapore, 2020; pp. 55–70. [CrossRef]
4. Borst, A.; Box, A.T.A.; Fluit, A.C. False-Positive Results and Contamination in Nucleic Acid Amplification Assays: Suggestions for a Prevent and Destroy Strategy. *Eur. J. Clin. Microbiol. Infect. Dis.* **2004**, *23*, 289–299. [CrossRef] [PubMed]
5. Fruehwirth, M.; Rivas, A.; Fitz, A.; Cechinel-Assing-Batista, A.; Silveira, C.; Delai, R. False-negative result in molecular diagnosis of SARS-CoV-2 in samples with amplification inhibitors. *J. Bras. Patol. Med. Lab.* **2020**, *56*, 1–8. [CrossRef]
6. Conklin, S.E.; Martin, K.; Manabe, Y.C.; Schmidt, H.A.; Miller, J.; Keruly, M.; Klock, E.; Kirby, C.S.; Baker, O.R.; Fernandez, R.E.; et al. Evaluation of Serological SARS-CoV-2 Lateral Flow Assays for Rapid Point-of-Care Testing. *J. Clin. Microbiol.* **2021**, *59*, e02020-20. [CrossRef]
7. Kilic, T.; Weissleder, R.; Lee, H. Molecular and Immunological Diagnostic Tests of COVID-19: Current Status and Challenges. *iScience* **2020**, *23*, 101406–101424. [CrossRef] [PubMed]
8. OECD. *Testing for COVID-19: How to Best Use the Various Tests? OECD Policy Responses to Coronavirus (COVID-19)*; OECD Publishing: Paris, France, 2020; Available online: https://www.oecd.org/coronavirus/policy-responses/testing-for-covid-19-how-to-best-use-the-various-tests-c76df201/ (accessed on 24 April 2021).
9. Orooji, Y.; Sohrabi, H.; Hemmat, N.; Oroojalian, F.; Baradaran, B.; Mokhtarzadeh, A.; Mohaghegh, M.; Karimi-Maleh, H. An Overview on SARS-CoV-2 (COVID-19) and Other Human Coronaviruses and Their Detection Capability via Amplification Assay, Chemical Sensing, Biosensing, Immunosensing, and Clinical Assays. *Nano-Micro Lett.* **2021**, *13*, 18. [CrossRef]
10. Li, M.; Wu, J.; Ma, M.; Feng, Z.; Mi, Z.; Rong, P.; Liu, D. Alkyne- and nitrile-anchored gold nanoparticles for multiplex SERS imaging of biomarkers in cancer cells and tissues. *Nanotheranostics* **2019**, *3*, 113–119. [CrossRef]
11. Nishi, K.; Isobe, S.; Zhu, Y.; Kiyama, R. Fluorescence-based bioassays for the detection and evaluation of food materials. *Sensors* **2015**, *15*, 25831–25867. [CrossRef]
12. Jiang, L.; Qian, J.; Cai, F.; He, S. Raman reporter-coated gold nanorods and their applications in multimodal optical imaging of cancer cells. *Anal Bioanal Chem.* **2011**, *400*, 2793–2800. [CrossRef]
13. Mugo, S.M.; Zhang, Q. Nano-Sized Structured Platforms for Facile Solid-Phase Nanoextraction for Molecular Capture and (Bio)Chemical Analysis. In *Micro and Nano Technologies, Nanomaterials Design for Sensing Applications*; Zenkina, O.V., Ed.; Elsevier: Amsterdam, The Netherlands, 2019; pp. 153–195. [CrossRef]

14. Egan, J.G.; Drossis, N.; Ebralidze, I.I.; Fruehwald, H.M.; Laschuk, N.O.; Poisson, J.; de Haan, H.W.; Zenkina, O.V. Hemoglobin-driven iron-directed assembly of gold nanoparticles. *RSC Adv.* **2018**, *8*, 15675–15686. [CrossRef]
15. Kim, S.M.; Kim, J.; Noh, S.; Sohn, H.; Lee, T. Recent Development of Aptasensor for Influenza Virus Detection. *BioChip J.* **2020**, *14*, 327–339. [CrossRef] [PubMed]
16. Negri, P.; Chen, G.; Kage, A.; Nitsche, A.; Naumann, D.; Xu, B.; Dluhy, R.A. Direct Optical Detection of Viral Nucleoprotein Binding to an Anti-Influenza Aptamer. *Anal. Chem.* **2012**, *84*, 5501–5508. [CrossRef] [PubMed]
17. Chen, H.; Park, S.-G.; Choi, N.; Moon, J.-I.; Dang, H.; Das, A.; Lee, S.; Kim, D.-G.; Chen, L.; Choo, J. SERS imaging-based aptasensor for ultrasensitive and reproducible detection of influenza virus A. *Biosens. Bioelectron.* **2020**, *167*, 112496. [CrossRef] [PubMed]
18. Stanborough, T.; Given, F.M.; Koch, B.; Sheen, C.R.; Stowers-Hull, A.B.; Waterland, M.R.; Crittenden, D.L. Optical Detection of CoV-SARS-2 Viral Proteins to Sub-Picomolar Concentrations. *ACS Omega* **2021**, *6*, 6404–6413. [CrossRef]
19. Yang, Y.; Peng, Y.; Lin, C.; Long, L.; Hu, J.; He, J.; Zeng, H.; Huang, Z.; Li, Z.Y.; Tanemura, M.; et al. Human ACE2-Functionalized Gold "Virus-Trap" Nanostructures for Accurate Capture of SARS-CoV-2 and Single-Virus SERS Detection. *Nanomicro Lett.* **2021**, *13*, 109. [CrossRef]
20. Kukushkin, V.I.; Ivanov, N.M.; Novoseltseva, A.A.; Gambaryan, A.S.; Yaminsky, I.V.; Kopylov, A.M.; Zavyalova, E.G. Highly sensitive detection of influenza virus with SERS aptasensor. *PLoS ONE.* **2019**, *14*, e0216247. [CrossRef]
21. Gribanyov, D.; Zhdanov, G.; Olenin, A.; Lisichkin, G.; Gambaryan, A.; Kukushkin, V.; Zavyalova, E. SERS-Based Colloidal Aptasensors for Quantitative Determination of Influenza Virus. *Int. J. Mol. Sci.* **2021**, *22*, 1842. [CrossRef]
22. Song, Y.; Song, J.; Wei, X.; Huang, M.; Sun, M.; Zhu, L.; Yang, C. Discovery of Aptamers Targeting Receptor-Binding Domain of the SARS-CoV-2 Spike Glycoprotein. *Anal. Chem.* **2020**, *92*, 9895–9900. [CrossRef]
23. Leopold, A.N.; Lendl, B. A New Method for Fast Preparation of Highly Surface-Enhanced Raman Scattering (SERS) Active Silver Colloids at Room Temperature by Reduction of Silver Nitrate with Hydroxylamine Hydrochloride. *J. Phys. Chem. B.* **2003**, *107*, 5723–5727. [CrossRef]
24. Reed, L.J.; Muench, H. A Simple Method of Estimating Fifty Per Cent Endpoint. *Am. J. Epidemiol.* **1938**, *27*, 493–497. [CrossRef]
25. SPR-Pages. Available online: https://www.sprpages.nl/ (accessed on 15 March 2021).
26. Kramberger, P.; Ciringer, M.; Štrancar, A.; Peterka, M. Evaluation of nanoparticle tracking analysis for total virus particle determination. *Virol. J.* **2012**, *9*, 265. [CrossRef]
27. The ViennaRNA Web Services. Available online: http://rna.tbi.univie.ac.at/ (accessed on 20 April 2021).
28. Mergny, J.-L.; Lacroix, L. Analysis of Thermal Melting Curves. In *Oligonucleotides*; Mary Ann Liebert, Inc.: New Rochelle, NY, USA, 2003; Volume 13, pp. 515–537. [CrossRef]
29. Kypr, J.; Kejnovska, I.; Renciuk, D.; Vorlickova, M. Circular dichroism and conformational polymorphism of DNA. *Nucl. Acids Res.* **2009**, *37*, 1713–1725. [CrossRef]
30. Villar-Guerra, R.; Trent, J.O.; Chaires, J.B. G-Quadruplex Secondary Structure Obtained from Circular Dichroism Spectroscopy. *Angew. Chem. Int. Ed.* **2018**, *57*, 7171–7175. [CrossRef]
31. Elechiguerra, J.L.; Burt, J.L.; Morones, J.R.; Camacho-Bragado, A.; Gao, X.; Lara, H.H.; Yacaman, M.J. Interaction of silver nanoparticles with HIV-1. *J. Nanobiotechnol.* **2005**, *3*, 6. [CrossRef]
32. Sepunarua, L.; Plowmana, B.J.; Sokolova, S.V.; Young, N.P.; Comptona, R.G. Rapid Electrochemical Detection of Single Influenza Viruses Tagged with Silver Nanoparticles. *Chem. Sci.* **2016**, *7*, 3892–3899. [CrossRef]
33. Betz, J.F.; Yu, W.W.; Cheng, Y.; White, I.M.; Rubloff, G.W. Simple SERS substrates: Powerful, portable, and full of potential. *Phys. Chem. Chem. Phys.* **2014**, *16*, 2224–2239. [CrossRef]
34. Kukushkin, V.I.; Astrakhantseva, A.S.; Morozova, E.N. Influence of the Morphology of Metal Nanoparticles Deposited on Surfaces of Silicon Oxide on the Optical Properties of SERS Substrates. *Bull. Russ. Acad. Sci. Phys.* **2021**, *85*, 133–140. [CrossRef]
35. Kukushkin, V.I.; Grishina, Y.V.; Solov'ev, V.V.; Kukushkin, I.V. Size plasmon–polariton resonance and its contribution to the giant enhancement of the Raman scattering. *JETP Lett.* **2017**, *105*, 677–681. [CrossRef]
36. Jesus Gonçalves Rubira, R.; Alessio Camacho, S.; Silva Martin, C.; Mejía-Salazar, J.R.; Reyes Gómez, F.; Rosa da Silva, R.; Novais de Oliveira, O., Jr.; Alessio, P.; José Leopoldo Constantino, C. Designing Silver Nanoparticles for Detecting Levodopa (3,4-Dihydroxyphenylalanine, L-Dopa) Using Surface-Enhanced Raman Scattering (SERS). *Sensors* **2020**, *20*, 15. [CrossRef]
37. Ma, X.; Xumin, X.X.; Xia, Y.; Wang, Z. SERS Aptasensor for Salmonella typhimurium Detection based on Spiny Gold Nanoparticles. *Food Control* **2018**, *84*, 232–237. [CrossRef]
38. Duana, N.; Shen, M.; Qi, S.; Wang, W.; Wu, S.; Wang, Z. A SERS aptasensor for simultaneous multiple pathogens detection using gold decorated PDMS substrate. *Spectrochim. Acta Part A Mol. Biomol. Spectrosc.* **2020**, *230*, 118103. [CrossRef]
39. Tu, D.; Garzaab, J.T.; Cote, G.L. A SERS aptasensor for sensitive and selective detection of bis(2-ethylhexyl)phthalate. *RSC Adv.* **2019**, *9*, 2618–2625. [CrossRef]
40. Yang, L.; Fu, C.; Wang, H.; Xu, S.; Xu, W. Aptamer-based surface-enhanced Raman scattering (SERS) sensor for thrombin based on supramolecular recognition, oriented assembly, and local field coupling. *Anal. Bioanal. Chem.* **2017**, *409*, 235–242. [CrossRef] [PubMed]
41. Fedotova, Y.V.; Kukushkin, V.I.; Solovyev, V.V.; Kukushkin, I.V. Spoof plasmons enable giant Raman scattering enhancement in Near-Infrared region. *Opt. Express* **2019**, *27*, 32578–32586. [CrossRef]

42. Petti, L.; Capasso, R.; Rippa, M.; Pannico, M.; La Manna, P.; Peluso, G.; Musto, P. A plasmonic nanostructure fabricated by electron beam lithography as a sensitive and highly homogeneous SERS substrate for bio-sensing applications. *Vib. Spectrosc.* **2016**, *82*, 22–30. [CrossRef]
43. Mahmoud, A.Y.; Rusin, C.J.; McDermott, M.T. Gold nanostars as a colloidal substrate for in-solution SERS measurements using a handheld Raman spectrometer. *Analyst* **2020**, *145*, 1396–1407. [CrossRef] [PubMed]

 *nanomaterials*

Article

Charge-Transfer Induced by the Oxygen Vacancy Defects in the Ag/MoO3 Composite System

Qi Chu [1], Jingmeng Li [2], Sila Jin [3], Shuang Guo [3], Eungyeong Park [3], Jiku Wang [1,*], Lei Chen [1,*] and Young Mee Jung [3,*]

1 College of Chemistry, Jilin Normal University, Siping 136000, China; cq325626@163.com
2 School of Public Health and Basic Medicine, The Chinese University of Hong Kong, Hong Kong 999077, China; lijingmengapp@163.com
3 Department of Chemistry, Institute for Molecular Science and Fusion Technology, Kangwon National University, Chunchon 24341, Korea; jsira@kangwon.ac.kr (S.J.); guoshuang@kangwon.ac.kr (S.G.); egpark@kangwon.ac.kr (E.P.)
* Correspondence: jikuwang@sina.com (J.W.); chenlei@jlnu.edu.cn (L.C.); ymjung@kangwon.ac.kr (Y.M.J.); Tel.: +82-33-250-8495 (Y.M.J.)

Abstract: In this paper, an Ag/MoO_3 composite system was cosputtered by Ar plasma bombardment on a polystyrene (PS) colloidal microsphere array. The MoO_3 formed by this method contained abundant oxygen vacancy defects, which provided a channel for charge transfer in the system and compensated for the wide band gap of MoO_3. Various characterization methods strongly demonstrated the existence of oxygen vacancy defects and detected the properties of oxygen vacancies. 4-Aminothiophenol (p-aminothiophenol, PATP) was used as a candidate surface-enhanced Raman scattering (SERS) probe molecule to evaluate the contribution of the oxygen vacancy defects in the Ag/MoO_3 composite system. Interestingly, oxygen vacancy defects are a kind of charge channel, and their powerful effect is fully reflected in their SERS spectra. Increasing the number of charge channels and increasing the utilization rate of the channels caused the frequency of SERS characteristic peaks to shift. This interesting phenomenon opens up a new horizon for the study of SERS in oxygen-containing semiconductors and provides a powerful reference for the study of PATP.

Keywords: SERS; oxygen vacancy defects; Ag/MoO_3 composite; charge-transfer

Citation: Chu, Q.; Li, J.; Jin, S.; Guo, S.; Park, E.; Wang, J.; Chen, L.; Jung, Y.M. Charge-Transfer Induced by the Oxygen Vacancy Defects in the Ag/MoO_3 Composite System. *Nanomaterials* **2021**, *11*, 1292. https://doi.org/10.3390/nano11051292

Academic Editors: Ronald Birke and Bing Zhao

Received: 24 April 2021
Accepted: 11 May 2021
Published: 14 May 2021

1. Introduction

Surface-enhanced Raman scattering (SERS) has attracted much attention in the detection of biomolecules, chemicals, composite materials, environmental pollutants, etc. [1–5] and is a simple, fast, and sensitive technology. SERS phenomena were first observed on the surface of noble metals such as Ag, Au, and Cu, and noble metals are the first-choice materials for the early preparation of SERS substrates [6,7]. In 1974, Fleischman et al. was initially discovered SERS phenomenon on the rough silver electrode. Then, Van Duyne, Albrecht, and Creighton et al. proposed that some of the enhancement in the Raman scattering signal of the probe molecule comes from the surface enhancement of the rough silver electrode, and named this effect as surface-enhanced Raman scattering [8–10]. Due to the poor compatibility of noble metal materials, the adsorption sites of molecules on the surface of traditional noble metal SERS active materials are not uniform, and adsorption denaturation and other problems occur [11]. When a laser irradiates a noble metal, strong plasmon resonance and plasmon-driven catalysis reactions are induced, resulting in the poor reproducibility of SERS spectra [12]. Recent studies have shown that laser-induced inelastic scattering occurs on the surface of semiconductor thin films, leading to changes in the magnetic anisotropy and other properties of the semiconductor surface. These studies have proposed a method to improve the performance of semiconductors using magnons [13,14]. Raman scattering is also a kind of inelastic scattering. When this effect occurs on the surface of a semiconductor film, it will also cause changes in the distribution of electrons on the surface of the semiconductor. In recent years, there have been many studies on semiconductors exhibiting SERS activity [15–17]. By controlling the geometric structure, the electron density, the crystal morphology, and other semiconductor parameters, the ability for semiconductors to recognize target molecules can be enhanced, leading to better SERS performance. However, the low carrier concentrations on semiconductor surfaces result in weak surface plasmon resonance (SPR) in the visible area. Therefore, the optimal SERS effect can be achieved by utilizing the noble metal-semiconductor system. By combining noble metals with semiconductors, the SERS effect can be optimized, as the noble metals contribute electromagnetic (EM) enhancement, and the chemical enhancement (CM) of the system is stimulated. This approach has developed into a mature SERS research method [18–20].

In recent years, there has been some discussion on composite substrates composed of, for example, noble/noble metals (Au/Ag) and semiconductors (TiO_2, ZnO, Cu_2O, etc.). Based on previous reports, possible charge transfer (CT) paths in various structures were obtained. In substrates composed of noble metals and semiconductors, there are several assemblies that can form in the presence of probe molecules: metal-semiconductor-probe molecule, metal-probe molecule-semiconductor, and semiconductor-metal-probe molecule [21–24]. Yang et al. employed the self-assembly method using Au and ZnO as a composite substrate, and 4-aminothiophenol (p-aminothiophenol, PATP) was selected as the probe molecule to obtain the Au-ZnO-PATP system. In this system, the addition of ZnO significantly improved the enhancement effect. It was also confirmed that CT in the system, that is, through ZnO, involved charge carriers moving from Au to PATP [23]. Ji et al., using p-mercaptothiophenol (MPH) as a probe molecule and the noble metal Ag and semiconductor TiO_2 as a composite substrate, assembled a Ag-MPH-TiO_2 system [25]. They found that Ag provided a strong EM contribution in the system and that the introduction of the semiconductor TiO_2 changed the CM proportion. The results showed that in the Ag-MPH system, charge was transferred from MPH to Ag, and electron transfer from Ag to MPH and from MPH to TiO_2 occurred after TiO_2 was introduced, resulting in the overall transition of electrons from Ag to TiO_2. Jiang et al. synthesized a TiO_2-p-mercaptobenzoic acid (MBA)-Ag structure. The results confirmed that in such a structure, TiO_2 generates a surface state energy level after being excited by a laser. This energy level caused CT between TiO_2 and MBA, and the introduction of Ag produced a relatively obvious EM contribution; as such, the SERS effect of the MBA molecule was significantly enhanced [26]. Therefore, semiconductors play an important role in the SERS of composite systems.

Molybdenum trioxide (MoO_3) has many unique properties and exhibits outstanding performance in electrochromic materials, photochromic materials, photocatalysis, and biosensing [26–29]. MoO_3 has also received much attention in the field of SERS [30–32]. Prabhu et al. synthesized MoO_3 with a sea urchin morphology and obtained improved SERS results. [31] Zhu et al. doped MoO_3 with hydrogen under mild conditions, and the carrier concentration on the surface of MoO_3 was able to reach the concentration level of noble metals. [30] In addition, MoO_3 with abundant oxygen vacancies and a specific structure has facilitated the adsorption of SERS probe molecules. Therefore, MoO_3 has very broad prospects in the SERS field.

Previously, we reported the Ag-ZnSe-PATP system, which is the combination of noble metals and semiconductors. In Ag-ZnSe-PATP system, Ag and ZnSe were layer-by-layer sputtered on the polystyrene (PS) template. The proposed Ag-ZnSe composites compensate the CT difficulty in wide band gap semiconductors, which was initiated by the SPR of Ag [33]. In this study, we designed a Ag-MoO_3-PATP system and performed various characterizations and SERS analyses. Ag and MoO_3 thin films were deposited on PS colloidal microspheres by magnetron sputtering under the bombardment of an Ar plasma. The structure and characteristics of MoO_3 formed by this method do not change significantly; however, it does contain abundant oxygen vacancies. This is because the high-energy Ar plasma selectively activates oxygen atoms, causing the oxygen atoms to diffuse from the bulk of the oxygen-containing semiconductor to its surface. Previous studies have confirmed that oxygen vacancies form when an Ar plasma bombards oxygen-containing semiconductor films. The effect of the obtained defect energy level on the generation of electrons and holes has been confirmed by theoretical calculations [34]. In our research, bombarding the Ag-MoO_3 system with an Ar plasma also produced a substrate rich in oxygen vacancies, which was confirmed by transmission electron microscopy (TEM) and X-ray photoelectron spectroscopy (XPS) results. The presence of oxygen vacancies in the Ag-MoO_3 system strengthens CT [16] between the two materials and is reflected in the SERS spectrum of PATP, resulting in band positions that have not been assigned in previous studies.

2. Materials and Methods

2.1. Materials

Ag targets (99.99%) were purchased from Beijing Hezong Tianrui Technology Development Center. MoO_3 (99.99%) target materials were purchased from Beijing Jingmaiyan Material Technology Co., Ltd. (Beijing, China). PS colloidal microspheres with 200 nm diameters were purchased from Sinopharm Chemical Reagent Co., Ltd (Shanghai, China). PATP was purchased from Aladdin Reagent Co., Ltd. (Shanghai, China). Anhydrous ethanol was purchased from Sinopharm Chemical Reagent Co., Ltd. (Shanghai, China). All chemical reagents were used without further purification.

2.2. Preparation of the Ag/MoO_3 Two-Dimensional Arrays

The preparation process for the Ag/MoO_3 two-dimensional arrays is shown in Scheme 1. The PS microspheres were diluted with an equal volume of absolute ethanol, and the diluted microspheres were dispersed on the surface of a washed silicon wafer (10 cm × 5 cm). The silicon wafer was then placed in water so that the surface of PS microspheres remained on the water surface. A second washed silicon wafer (2 cm × 2 cm) were placed in the water and were gently picked up to obtain the monolayer of PS microspheres. The PS microspheres on the water surface were removed to obtain a uniform microsphere film, and the film was dried in air for later use. In the magnetron sputtering system (ATC 1800-F, AJA, Scituate, MA, USA), the target materials were bombarded with Ar gas, and Ag and MoO_3 were simultaneously sputtered on the prepared PS array. The Ag sputtering power was 10 W, the MoO_3 sputtering power was 50, 70, or 90 W, and the sputtering time was 15 min. In addition, a sample was prepared by only sputtering Ag, and the sputtering power and sputtering time were 10 W and 15 min, respectively. A sample that was only sputtered with MoO_3 was also prepared, and the sputtering power and sputtering time were 90 W and 15 min, respectively. The working pressure of the system was 6×10^{-1} Pa, and the Ar flow rate was 9 sccm (standard cubic centimeters per minute). For SERS measurements, PATP was dissolved in absolute ethanol to prepare a solution with a concentration of 10^{-3} M. All samples were soaked with the PATP solution for 5 h and then washed with absolute ethanol three times.

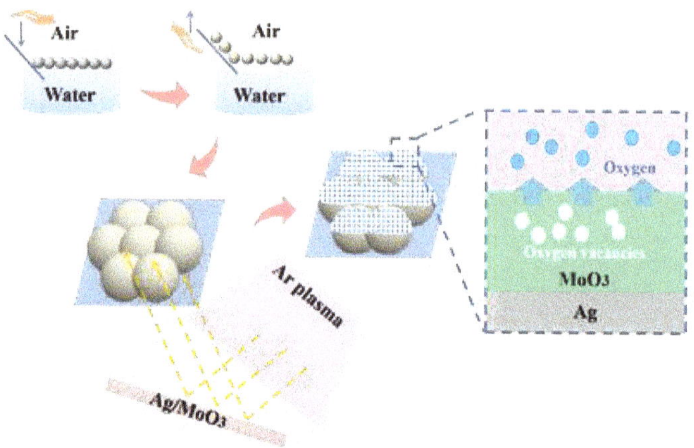

Scheme 1. Preparation of the Ag/MoO_3-coated PS template using the cosputtering technique.

2.3. Characterization of the Ag/MoO_3 Two-Dimensional Arrays

The morphologies of the different Ag/MoO_3 samples were observed with scanning electron microscopy (SEM, JEOL 6500F, JEOL, Tokyo, Japan) using a microscope operated at a voltage of 200 kV. Transmission electron microscopy (TEM, JEM-2100F, JEOL, Tokyo, Japan) was used to obtain high-resolution images of the samples under a 200 kV voltage. Ultraviolet-visible-near infrared (UV-Vis-NIR) absorption spectra were obtained using a Shimadzu UV-3600 spectrophotometer (Shimadzu, Kyoto, Japan). A Raman spectrometer (Renishaw Raman System Confocal 2000 Microphotometer, Renishaw, London, UK) with a 514 nm excitation light source equipped with a CCD detector and a holographic notch filter was used to obtain SERS spectra. The laser was focused on the surface of the sample through a 50× long-distance objective lens with a 1 μm spot size. When the SERS experiment was performed, the laser power employed was controlled from 100 to 20 mW. PATP is used as a probe molecule and is formulated into an absolute ethanol solution with a concentration of 10^{-3} M. The collection time was 20 s, and the number of acquisitions was 1. To determine the crystal structures of the samples, each sample was characterized by X-ray diffraction (XRD) using a Rigaku-MiniFlex600 system (Rigaku, Tokyo, Japan). To determine the element and its state, X-ray photoelectron spectroscopy (XPS) spectra of the sample were obtained with a Thermo-Scientific-Escalab 250 XI Al440 system (Thermo Fisher Scientific, Waltham, MA, USA), and all results were corrected with carbon peaks (C 1s = 284.6 eV).

3. Results

3.1. Characterization of the Ag/MoO_3-Coated PS Templates

As shown in Scheme 1, the self-assembly method was employed to prepare the PS microsphere arrays. The substrate was prepared by cosputtering Ag and MoO_3. The topography of the prepared substrates with different doping ratios can be visually observed. The morphology of Ag and MoO_3 on the cosputtered substrates was characterized by using SEM. Figure 1 shows the morphology of the Ag/MoO_3 composites with different doping ratios. Figure 1a shows the sample sputtered with Ag, and the sputtering power and time were 10 W and 15 min, respectively. In the samples shown in Figure 1b–d, the Ag sputtering power was 10 W, the MoO_3 sputtering power was 50, 70, or 90 W, and the sputtering time was 15 min. Figure 1e shows the PS array sputtered with MoO_3, and the sputtering power and time were 90 W and 15 min, respectively. A PS array without sputtered Ag or MoO_3 is shown in Figure 1f for comparison. The elemental distribution of the samples is shown on the right side of the SEM images. With increasing MoO_3 sputtering power from 50 to 90 W, the gaps between the PS microspheres are gradually filled, and the surface of the sample tends to be smooth, which was confirmed by the TEM images in the upper right corner of each image. The changes in morphology observed in Figure 1 will affect the hot spot distribution of the substrate, thus affecting SERS characteristics [35].

Figure 1. SEM images of (**a**) sputtered Ag on the PS array and (**b–d**) cosputtered Ag and MoO_3 on the PS array with different MoO_3 sputtering powers: (**b**) 50 W, (**c**) 70 W and (**d**) 90 W. SEM images of (**e**) sputtered MoO_3 on the PS array and (**f**) the noncoated PS microsphere array. The element symbols provided in red indicate that the element is not included in the sample. The upper right corner of the image provides a TEM image of a single PS microsphere.

Figure 2 shows high-resolution TEM images of the Ag/MoO_3 composites. Figure 2a,b show that Ag and MoO_3 form a hemispherical core-shell on the PS microspheres. Figure 2c shows the crystallization of Ag on the substrate, and the lattice spacing is 0.233 nm, which is assigned to the (220) plane of Ag. Figure 2d shows the crystallization of MoO_3, and the lattice spacing is 0.231 nm, which is assigned to the (202) crystal plane of MoO_3. Figure 2 shows that in the Ag/MoO_3 system, both Ag and MoO_3 crystal lattices exist. The distribution of Ag and MoO_3 proves the uniformity of the system obtained by cosputtering. This method allows Ag and MoO_3 particles to be embedded alternately without covering each other and has little interference with SERS performance. In addition, oxygen vacancy defects also exist on the surface of the system, which is very beneficial to the charge transition. Interestingly, in Figure 2d, we also see an amorphous region (the region between the two red lines) between the two lattice planes. Previously, researchers identified and proved that this corresponds to oxygen vacancy defects [36]. We speculate that the amorphous regions appearing in Figure 2d also correspond to oxygen vacancy defects, and XPS is used to prove this speculation by analyzing the elemental states.

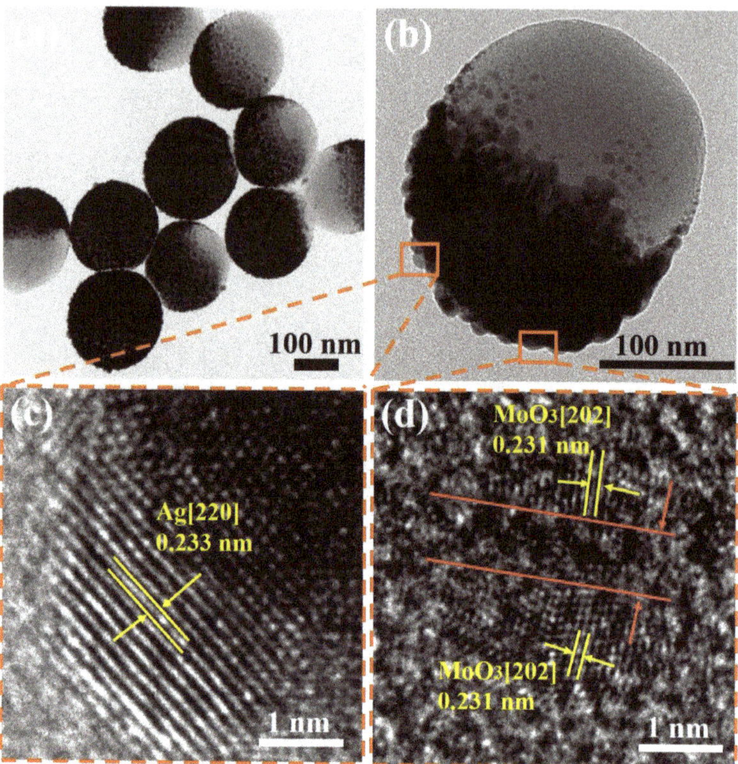

Figure 2. TEM images of the PS microspheres coated with Ag/MoO_3. SEM images showing (**a**) the aggregation of multiple Ag/MoO_3-coated microspheres and (**b**) a single Ag/MoO_3-coated microsphere. The lattice spacings for (**c**) Ag and (**d**) MoO_3 are 0.233 and 0.231 nm, respectively. The red lines indicate the corresponding oxygen vacancy defects.

The crystallization of the samples was determined from XRD spectra (Figure 3). Spectra a, b–d, and e in Figure 3 show the XRD spectra of the PS arrays sputtered with pure Ag, Ag-MoO_3 using different MoO_3 sputtering powers, and pure MoO_3, respectively. With increasing MoO_3 sputtering power, the intensity of the Ag diffraction peak decreases gradually. The diffraction peaks at 13.73°, 16.74°, 25.53°, 38.20°, 44.47°, 64.78°, and 77.82° correspond to the (110), (040), (111), (200), (220), (440), and (222) lattice planes of Ag, respectively. The diffraction peak at 69.39° corresponds to the (202) lattice plane of MoO_3. The diffraction peak for MoO_3 at 69.39° is sharp and strong, which indicates that the crystal is dominated by the (202) lattice plane. There is an impurity peak at 33.04° in spectrum e (Figure 3), which may be attributed to the amorphous region caused by oxygen vacancy defects. The XRD spectra show that with increasing MoO_3 content, the intensity and sharpness of the Ag diffraction peaks on the sample surface become weaker. This is because Ag becomes increasingly covered by MoO_3. The increase in MoO_3 content also leads to the SPR of Ag weakening, which is consistent with the SERS results.

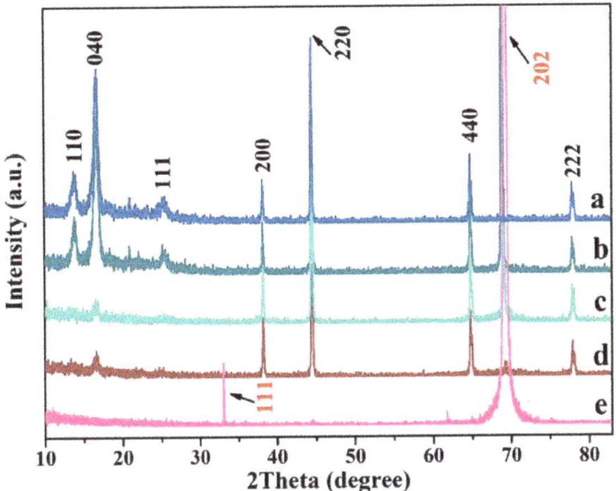

Figure 3. XRD spectra of the PS arrays sputtered with (**a**) Ag, (**b–d**) Ag/MoO_3, and (**e**) MoO_3. The MoO_3 sputtering powers were (**b**) 50, (**c**) 70 and (**d**) 90 W. The diffraction peaks annotated in black and red correspond to the lattice planes of Ag and MoO_3, respectively.

Figure 4 shows the XPS spectra of the Ag/MoO_3 system. The spectrum shows the chemical composition and electronic structure of each sample. From the XPS spectra of Ag and Mo elements shown in Figure 4A,B, respectively, we found that the characteristic peaks of the elements of the composite material shifted compared with those of the individual elements. This is due to the polarization of the electron cloud in the composite system, and the binding energy of the elements has changed. Figure 4C shows two states of oxygen through peak fitting. The O_a peak marked by the red dashed line comes from the oxygen vacancies in the matrix [37]. We found that with the increase in MoO_3 content, the peak attributed to oxygen vacancies gradually became obvious, and its intensity increased. The O_b peak comes from the lattice oxygen in MoO_3, and its intensity gradually weakens. From the XPS spectrum of oxygen, we confirmed the existence of oxygen vacancies and found changes in the content of oxygen vacancies. The results obtained by XPS analysis clearly confirmed that the amorphous region in the TEM image was indeed caused by oxygen vacancies. In addition, we calculated the percentage of the elemental content in each sample based on the XPS test results (as shown in Figure 5). In the calculation, we employed the carbon peak for correction to deduce the interference of carbon element. With the increase in MoO_3 content, we observed that the Ag content gradually decreases. Since the stoichiometric ratio of O to Mo in MoO_3 is 3:1, the O content increases significantly, while the Mo content increases slightly.

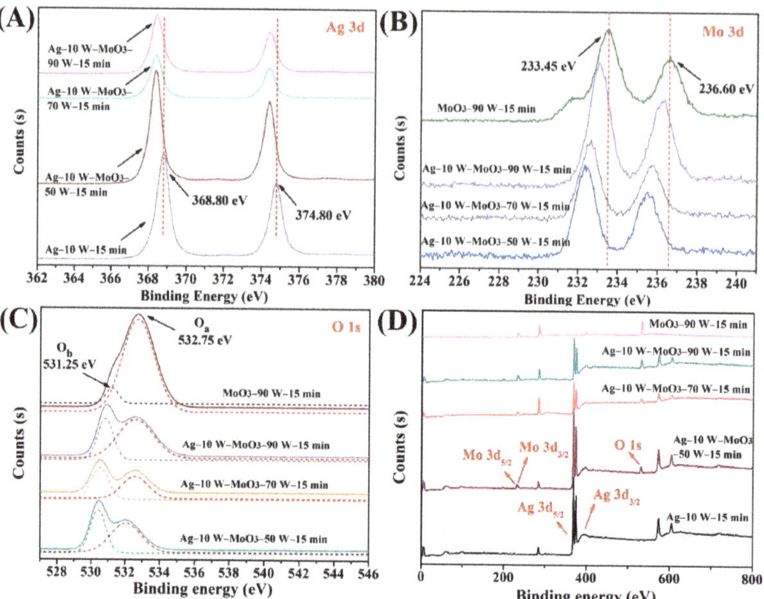

Figure 4. XPS spectra of the Ag/MoO_3 system. (**A–C**) are the spectra of Ag, Mo, and O, respectively. (**D**) is the survey spectra of Ag deposited with a sputtering power of 10 W and MoO_3 deposited with a sputtering power of 90 W, and for the Ag/MoO_3 system, the sputtering power for Ag was 10 W, and that for MoO_3 was 50, 70, and 90 W.

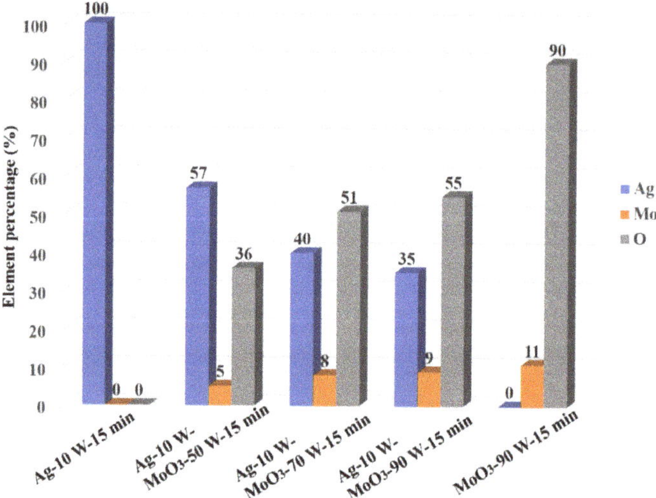

Figure 5. A histogram of the content of each element based on the XPS test results.

Figure 6 shows the UV-Vis-NIR absorption spectra of the PS arrays sputtered with Ag, Ag/MoO_3 and MoO_3. The peak attributed to PS microspheres appears at approximately 340 nm. With the increase in MoO_3 sputtering power, the content of Ag and MoO_3 on the surface of the hemispherical shell is constantly changing, and the distribution of each component is irregular. Therefore, the peak attributed to PS microspheres may be coupled with the Ag peak, thus changing the intensity and position of this peak. For the spectrum a in Figure 6, the spectrum of the PS array sputtered with Ag shows a broad band at 448 nm, which is attributed to the Ag shells on the PS microsphere array. A broad absorption band corresponding to MoO_3 appears at about 400–700 nm (spectrum e in Figure 6) [38]. With increasing MoO_3 sputtering power, the absorption bands for MoO_3 and Ag become coupled, leading to an increase in the band intensity. Additionally, a redshift is observed for the band assigned to the Ag shells (spectra b–d in Figure 6). Using PS microspheres with a diameter of 200 nm as a template, a hemisphere covered by Ag and MoO_3 was formed alternately. The intensity of the dipole at the edge of this hemispherical structure is very large, even greater than that of the top dipole [39]. Such a hemisphere with an increasingly stronger electric field from the top to the edge is formed. The electron group near the spherical shell enhances not only the SERS signal but also the localized surface plasmon resonance (LSPR) of the composite substrate (see the red asterisk). By increasing the MoO_3 content, a redshift is observed for the LSPR bands of the Ag/MoO_3 composites, as shown by the red asterisk in spectra b–d (Figure 6). Since the increase in the number of oxygen vacancy defects is a microscopic and gradual change, the red shift in this band is not obvious. Based on previous reports on MoO_3 [37,40,41], this is due to the existence of oxygen vacancy defects and their interaction with Ag. With an increase in the MoO_3 content, the oxygen vacancy defects reduce the actual band gap of MoO_3, CT between Ag and MoO_3 occurs easily, and the photon energy required by the system decreases. The redshift of the absorption band at approximately 500 nm is dependent on the MoO_3 sputtering power. The UV-Vis-NIR absorption spectra show changes in the optical properties, especially the LSPR. In addition, we accurately determined the excitation wavelength used in the SERS study, which was based on the position of the main absorption band in the UV-Vis-NIR absorption spectra.

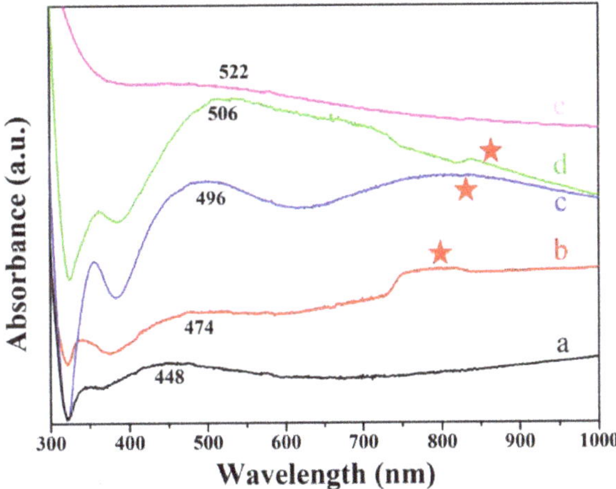

Figure 6. UV-Vis-NIR absorption spectra of the PS arrays sputtered with (**a**) Ag, (**b–d**) Ag/MoO_3 prepared using different MoO_3 sputtering powers, and (**e**) MoO_3. The MoO_3 sputtering powers were (**b**) 50, (**c**) 70 and (**d**) 90 W. The MoO_3 sputtering power for the sample described in (**e**) was 90 W. All the samples were sputtered for 15 min.

3.2. SERS Characteristics of the Ag/MoO_3-Coated PS Templates

PATP was employed as a probe molecule to evaluate the SERS properties of Ag/MoO_3. The samples were soaked in a PATP solution with a concentration of 10^{-3} M to complete binding to PATP. SERS spectra of PATP adsorbed on the Ag, Ag/MoO_3, and MoO_3-coated PS arrays are shown in Figure 7A. The laser power used at this time was 20 mW. The SERS spectra of PATP adsorbed on the Ag/MoO_3 composites exhibited unique changes when compared with the SERS spectrum of PATP adsorbed on the Ag-coated PS array. Two new obvious bands emerged at 1556 and 1168 cm^{-1} when the MoO_3 sputtering power was increased. In addition, a blueshift in the band at 1305 cm^{-1} was detected. This occurred due to the CT contribution from the Ag/MoO_3 composites. Therefore, the contribution of CT to SERS intensity was studied according to the equation used to describe the degree of CT (ρ_{CT}), which was proposed by Lombardi and Birke [42]:

$$\rho_{CT}(k) = [I_k(CT) - I_k(SPR)]/[I_k(CT) + I_0(SPR)]$$

where k corresponds to a single molecule in the SERS spectrum, I_k (CT) corresponds to the Raman band enhanced by CT, and I_0 (SPR) corresponds to the Raman band enhanced by EM. If $\rho_{CT}(k)$ is greater than 0.5, CT is mainly responsible for the SERS enhancement [41].

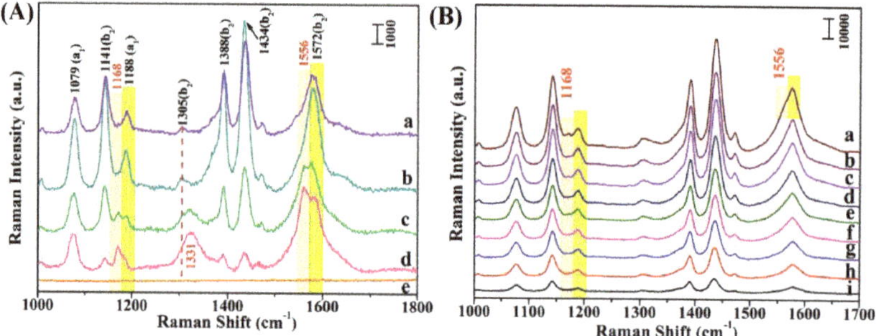

Figure 7. (**A**) SERS spectra of PATP adsorbed on (**a**) Ag, (**b–d**) Ag/MoO_3 prepared with different MoO_3 sputtering powers, and (**e**) MoO_3 coated on the PS arrays under 514 nm laser excitation. The MoO_3 sputtering powers were (**b**) 50, (**c**) 70 and (**d**) 90 W. (**B**) Power-dependent SERS spectra of PATP adsorbed on the Ag/MoO_3 composite (the sputtering powers for Ag and MoO_3 were 10 and 50 W). The laser powers used for the samples depicted in a–i were 100–20 mW, and the power difference was 10 mW.

As shown in Figure 7A, the strongest SERS signal is observed when the MoO_3 sputtering power is 50 W, which is the most conducive to SERS enhancement. Therefore, the ρ_{CT} equation was employed to evaluate the CT process. The ρ_{CT} of each sample was calculated with respect to the bands at 1434 and 1079 cm^{-1}. The ρ_{CT} values for the Ag/MoO_3 composites prepared with different MoO_3 sputtering powers are shown in Figure 7A.

As shown in the Figure 8A of the relationship between ρ_{CT} and the MoO_3 sputtering power, the ρ_{CT} values for the Ag/MoO_3 composites prepared with MoO_3 sputtering powers of 0, 50, 70, and 90 W are 0.48, 0.70, 0.64, and 0.53, respectively. The observation of two new bands and a blueshifted band is due to CT between the Ag/MoO_3 composites and PATP. The high ρ_{CT} is due to the oxygen vacancy defects in MoO_3, which provide an electron channel, and the electrons are excited by the SPR of Ag. Through this new channel, the electrons jump to the energy level of PATP, leading to an electron cloud rearrangement and the formation of a new b_2 mode. Although the structure of PATP is relatively simple, the physical and chemical activities of PATP are complex [43,44].

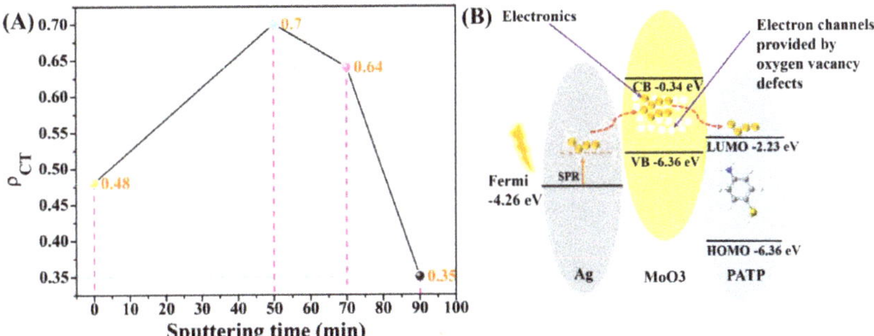

Figure 8. (**A**) The relationship between ρ_{CT} and the MoO_3 sputtering power. (**B**) The mechanism underlying CT between the Ag/MoO_3 composites and PATP.

With increasing MoO_3 sputtering power, the bands at 1168, 1331, and 1556 cm^{-1} become increasingly obvious. Combined with previously reported theoretical calculations on PATP [45], the band at 1168 cm^{-1} is attributed to the contribution of β(CH). The bands at 1331 and 1556 cm^{-1} are attributed to the contributions of β(CH) and υ(CC), respectively. The detailed band assignments are listed in Table 1. These new vibration modes are all related to the plane of the benzene ring. We believe that the abundant oxygen vacancy defects provide channels for electron movement. The large-scale "Ag/MoO_3-PATP" electron migration not only results in the vibration of chemical bonds outside the benzene ring but also causes the chemical bonds in the plane of the benzene ring to vibrate. Due to the relatively stable structure of the benzene ring, this additional vibration can only occur when CT is very active in the system. With increasing MoO_3 sputtering power, although the overall intensity of the SERS bands decreases, the positions of the bands at 1168, 1331, and 1556 cm^{-1} remain obvious, and their relative intensities increase, which confirms the effect of oxygen vacancy defects on the CT system. The CT mechanism is depicted in Figure 8B.

Table 1. Assignments for the Raman and SERS bands of and PATP adsorbed on the Ag/MoO_3-coated PS template [45].

Raman Shift/cm^{-1}	Band Assignments
1079	υ(CS) + υ(CS), 7a(a_1)
1141	β(CH), 9b(b_2)
1168	β(CH), 9a(a_1)
1188	β(CH), 9a(a_1)
1305	υ(CC), 15(b_2)
1331	β(CH), 14(b_2)
1388	β(CH), 14(b_2)
1434	υ(CC) + β(CH), 19b(b_2)
1556	υ(CC), 8(b_2)
1572	υ(CC), 8(b_2)

υ, stretching; β, bending. For ring vibrations, the corresponding vibrational modes of benzene and the symmetry species under $C_{2υ}$ symmetry are indicated. Where a_1 and b_2 represent totally and nontotally symmetric vibrational modes of the molecules, respectively.

To verify the importance of oxygen vacancy defects, the laser power-dependent experiment was performed. By increasing of the laser power from 20 to 100 mW (spectra a to i in Figure 7B), the intensities of bands at 1168 and 1556 cm^{-1} were significantly increased which indicated that the number of excited electrons was increased. Herein, the oxygen vacancy defect acts as an intermediate energy level (electron sink and the shallow donor level) and promotes the CT to the molecule (as shown in Figure 8B). Therefore, oxygen vacancy defect results in a broader donor energy level distribution, which provides the more effective passageways for electron transitions and compensates for the wide band gap of the MoO_3 semiconductor [33]. Thus, the higher laser power, the more excited the electrons; the higher utilization rate of oxygen vacancy defects, the larger the scale of electronic transitions.

4. Conclusions

We successfully prepared a Ag/MoO_3 composite system containing oxygen vacancy defects and performed a SERS study with PATP molecules. High-resolution TEM image analysis proved the existence of oxygen vacancy defects. In this system, the oxygen vacancy defects act as a charge channel to assist CT in wide band gap semiconductors. The SERS results show that the CT induced by oxygen vacancies results in the formation of a new b_2 mode band, which confirms our proposed mechanism. This study provides a reference for future studies of SERS in oxygen-containing semiconductors and new band assignments for PATP. Therefore, it opens a new field for the SERS-based study of oxygen vacancy defect-containing semiconductors.

Author Contributions: Writing—original draft preparation: Q.C.; data curation: J.L., S.J., S.G., E.P.; methodology: J.W.; writing—review and editing: L.C. and Y.M.J. All authors have read and agreed to the published version of the manuscript.

Funding: This research was funded by Jilin Normal University (Jishi Scholar). This work was also supported by the National Research Foundation of Korea (NRF) grants funded by the Korea government (no. NRF-2021R1A2C2004550, no. NRF-2020K2A9A2A06036299, and no. NRF-2020R1A4A1016093).

Institutional Review Board Statement: Not applicable.

Informed Consent Statement: Not applicable.

Data Availability Statement: Do not have the supporting reported results.

Conflicts of Interest: The authors declare no conflict of interest.

References

1. Stiles, P.L.; Dieringer, J.A.; Shah, N.C.; Van Duyne, R.P. Surface-enhanced Raman spectroscopy. *Anal. Chem.* **2008**, *1*, 601–626. [CrossRef]
2. Amendola, V.; Pilot, R.; Frasconi, M.; Marago, O.M.; Iati, M.A. Surface plasmon resonance in gold nanoparticles: A review. *J. Phys. Conden. Mat.* **2017**, *29*, 203002. [CrossRef]
3. Ding, S.Y.; You, E.M.; Tian, Z.Q.; Moskovits, M. Electromagnetic theories of surface-enhanced Raman spectroscopy. *Chem. Soc. Rev.* **2017**, *46*, 4042–4076. [CrossRef]
4. Han, B.; Guo, S.; Jin, S.; Park, E.; Xue, X.; Chen, L. Jung, Y.M. Improved charge transfer contribution by cosputtering Ag and ZnO. *Nanomaterials* **2020**, *10*, 1455. [CrossRef]
5. Bernatová, S.; Donato, M.G.; Ježek, J.; Pilát, Z.; Samek, O.; Magazzù, A.; Maragò, O.M.; Zemánek, P.; Gucciardi, P.G. Wavelength-dependent optical force aggregation of gold nanorods for SERS in a microfluidic chip. *J. Phys. Chem. C* **2019**, *9*, 5608–5615. [CrossRef]
6. Yan, X.; Wang, M.; Sun, X.; Wang, Y.; Shi, G.; Ma, W.; Hou, P. Sandwich-like Ag@Cu@CW SERS substrate with tunable nanogaps and component based on the Plasmonic nanonodule structures for sensitive detection crystal violet and 4-aminothiophenol. *Appl. Surf. Sci.* **2019**, *479*, 879–886. [CrossRef]
7. He, L.; Liu, C.; Tang, J.; Zhou, Y.; Yang, H.; Liu, R.; Hu, J. Self-catalytic stabilized Ag-Cu nanoparticles with tailored SERS response for plasmonic photocatalysis. *Appl. Surf. Sci.* **2018**, *434*, 265–272. [CrossRef]
8. Fleischman, M.; Hendra, P.J.; McQuillan, A.J. Raman spectra of pyridine adsorbed at a silver electrode. *Chem. Phys. Lett.* **1974**, *26*, 163–166. [CrossRef]
9. Jeanmarie, D.C.; Van Duyne, R.P. Surface raman spectroelectrochemistry: Part I. heterocyclic, aromatic, and aliphatic amines adsorbed on the anodized silver electrode. *J. Electroanal. Chem.* **1977**, *84*, 1–20. [CrossRef]
10. Albrecht, M.G.; Creighton, J.A. Anomalously intense raman spectra of pyridine at a silver electrode. *J. Am. Chem. Soc.* **1977**, *99*, 5215–5217. [CrossRef]
11. Sun, M.; Xu, H. A novel application of plasmonics: Plasmon-driven surface-catalyzed reactions. *Small Mol.* **2012**, *18*, 2777–2786. [CrossRef] [PubMed]
12. Lei, C.; Cai, L.; Ruan, W.; Bing, Z. Surface-enhanced Raman spectroscopy (SERS): Protein application. In *Encyclopedia of Analytical Chemistry*; John Wiley & Sons Ltd.: Chichester, UK, 2014.

13. Sadovnikov, A.V.; Gubanov, V.A.; Sheshukova, S.E.; Sharaevskii, Y.P.; Nikitov, S.A. Spin-wave drop filter based on asymmetric side-coupled magnonic crystals. *Phys. Rev. Appl.* **2018**, *9*, 051002. [CrossRef]
14. Sadovnikov, A.V.; Beginin, E.N.; Sheshukova, S.E.; Sharaevskii, Y.P.; Stognij, A.I.; Novitski, N.N.; Sakharov, V.K.; Khivintsev, Y.V.; Nikitov, S.A. Route toward semiconductor magnonics: Light-induced spin-wave nonreciprocity in a YIG/GaAs structure. *Phys. Rev. B* **2019**, *99*, 054424. [CrossRef]
15. Lin, J.; Hao, W.; Shang, Y.; Wang, X.; Qiu, D.; Ma, G.; Chen, C.; Li, S.; Guo, L. Direct experimental observation of facet-dependent SERS of Cu_2O polyhedra. *Small Mol.* **2018**, *14*, 8.
16. Wu, H.; Wang, H.; Li, G. Metal oxide semiconductor SERS-active substrates by defect engineering. *Analyst* **2017**, *142*, 326–335. [CrossRef] [PubMed]
17. Yang, J.L.; Xu, J.; Ren, H.; Sun, L.; Xu, Q.C.; Zhang, H.; Li, J.F.; Tian, Z.Q. In situ SERS study of surface plasmon resonance enhanced photocatalytic reactions using bifunctional Au@CdS core-shell nanocomposites. *Nanoscale Res. Lett.* **2017**, *9*, 6254–6258. [CrossRef]
18. Mitsai, E.; Kuchmizhak, A.; Pustovalov, E.; Sergeev, A.; Mironenko, A.; Bratskaya, S.; Linklater, D.P.; Balcytis, A.; Ivanova, E.; Juodkazis, S. Chemically non-perturbing SERS detection of a catalytic reaction with black silicon. *Nanoscale Res. Lett.* **2018**, *10*, 9780–9787. [CrossRef]
19. Zhang, X.Y.; Han, D.; Pang, Z.; Sun, Y.; Wang, Y.; Zhang, Y.; Yang, J.; Chen, L. Charge transfer in an ordered Ag/Cu_2S/4-MBA system based on surface-enhanced raman scattering. *J. Phys. Chem. C* **2018**, *122*, 5599–5605. [CrossRef]
20. Su, S.; Zhang, C.; Yuwen, L.; Chao, J.; Zuo, X.; Liu, X.; Song, C.; Fan, C.; Wang, L. Creating SERS hot spots on MoS_2 nanosheets with in situ grown gold nanoparticles. *ACS Appl. Mater. Interfaces* **2014**, *6*, 18735–18741. [CrossRef]
21. Ye, J.; Hutchison, J.A.; Uji-i, H.; Hofkens, J.; Lagae, L.; Maes, G.; Borghs, G.; Van Dorpe, P. Excitation wavelength dependent surface enhanced Raman scattering of 4-aminothiophenol on gold nanorings. *Nanoscale Res. Lett.* **2012**, *4*, 1606–1611. [CrossRef] [PubMed]
22. Dutta, S.; Ray, C.; Sarkar, S.; Pradhan, M.; Negishi, Y.; Pal, T. Silver nanoparticle decorated reduced graphene oxide (rGO) nanosheet: A platform for SERS based low-level detection of uranyl ion. *ACS Appl. Mater. Interfaces* **2013**, *5*, 8724–8732. [CrossRef] [PubMed]
23. Liu, L.; Yang, H.; Ren, X.; Tang, J.; Li, Y.; Zhang, X.; Cheng, Z. Au-ZnO hybrid nanoparticles exhibiting strong charge-transfer-induced SERS for recyclable SERS-active substrates. *Nanoscale Res. Lett.* **2015**, *7*, 5147–5151. [CrossRef] [PubMed]
24. Yang, L.; Sang, Q.; Du, J. A Ag synchronously deposited and doped TiO_2 hybrid as an ultrasensitive SERS substrate: A multifunctional platform for SERS detection and photocatalytic degradation. *Phys. Chem. Chem. Phys.* **2018**, *20*, 15149–15157. [CrossRef] [PubMed]
25. Ji, W.; Kitahama, Y.; Xue, X. Generation of pronounced resonance profile of charge-transfer contributions to surface-enhanced raman scattering. *J. Phys. Chem. C* **2012**, *116*, 2515–2520. [CrossRef]
26. Jiang, X.; Sun, X.; Yin, D. Recyclable Au-TiO_2 nanocomposite SERS-active substrates contributed by synergistic charge transfer effect. *Phys. Chem. Chem. Phys.* **2017**, *19*, 11212–11219. [CrossRef] [PubMed]
27. Chen, Z.; Cummins, D.; Reinecke, B.N.; Clark, E.; Sunkara, M.K.; Jaramillo, T.F. Core-shell MoO_3-MoS_2 nanowires for hydrogen evolution: A functional design for electrocatalytic materials. *Nano Lett.* **2011**, *11*, 4168–4175. [CrossRef]
28. Kim, H.S.; Cook, J.B.; Lin, H.; Ko, J.S.; Tolbert, S.H.; Ozolins, V.; Dunn, B. Oxygen vacancies enhance pseudocapacitive charge storage properties of MoO_{3-x}. *Nat. Mater.* **2017**, *16*, 454–460. [CrossRef]
29. Lee, Y.J.; Seo, Y.I.; Kim, S.H.; Kim, D.G.; Kim, Y.D. Optical properties of molybdenum oxide thin films deposited by chemical vapor transport of $MoO_3(OH)_2$. *Appl. Phys. A* **2009**, *97*, 237–241. [CrossRef]
30. Zhu, Q.; Jiang, S.; Ye, K.; Hu, W.; Zhang, J.; Niu, X.; Lin, Y.; Chen, S.; Song, L.; Zhang, Q.; et al. Hydrogen-doping-induced metal-like ultrahigh free-carrier concentration in metal-oxide material for giant and tunable plasmon resonance. *Adv. Mater.* **2020**, *50*, 2004059. [CrossRef]
31. Prabhu, B.R.; Bramhaiah, K.; Singh, K.K.; John, N.S. Single sea urchin–MoO_3 nanostructure for surface enhanced Raman spectroscopy of dyes. *Nanoscale* **2019**, *1*, 2426–2434. [CrossRef]
32. Shi, T.; Liang, P.; Zhang, X.; Zhang, D.; Shu, H.; Huang, J.; Yu, Z.; Xu, Y. Synergistic enhancement effect of MoO_3@Ag hybrid nanostructures for boosting selective detection sensitivity. *Spectrochim. Acta A Mol. Biomol. Spectrosc.* **2020**, *241*, 118611. [CrossRef]
33. Chu, Q.; Han, B.; Jin, Y.; Guo, S.; Jin, S.; Park, E.; Chen, L.; Jung, Y.M. Surface plasmon resonance induced charge transfer effect on the Ag-ZnSe-PATP system. *Spectrochim. Acta A Mol. Biomol. Spectrosc.* **2021**, *248*, 119167. [CrossRef]
34. Yang, L.; Yin, D.; Shen, Y.; Yang, M.; Li, X.; Han, X.; Jiang, X.; Zhao, B. Highly-dispersed TiO_2 nanoparticles with abundant active sites induced by surfactants as a prominent substrate for SERS: Charge transfer contribution. *Phys. Chem. Chem. Phys.* **2017**, *19*, 22302–22308. [CrossRef]
35. Ding, S.Y.; Yi, J.; Li, J.F.; Ren, B.; Wu, D.Y.; Panneerselvam, R.; Tian, Z.Q. Nanostructure-based plasmon-enhanced Raman spectroscopy for surface analysis of materials. *Nat. Rev. Mater.* **2016**, *1*, 6. [CrossRef]
36. Yan, X.; Li, Y.; Zhao, J.; Li, Y.; Bai, G.; Zhu, S. Roles of grain boundary and oxygen vacancies in $Ba_{0.6}Sr_{0.4}TiO_3$ films for resistive switching device application. *Appl. Phys. Lett.* **2016**, *108*, 033108. [CrossRef]
37. Mao, Y.; Li, W.; Sun, X.; Ma, Y.; Xia, J.; Zhao, Y.; Lu, X.; Gan, J.; Liu, Z.; Chen, J.; et al. Room-temperature ferromagnetism in hierarchically branched MoO_3 nanostructures. *CrystEngComm* **2012**, *4*, 1419–1424. [CrossRef]

38. Niu, Z.; Zhou, C.; Wang, J.; Xu, Y.; Gu, C.; Jiang, T.; Zeng, S.; Zhang, Y.; Ang, D.S.; Zhou, J. UV-light-assisted preparation of MoO_{3-x}/Ag NPs film and investigation on the SERS performance. *Asian J. Mater. Sci.* **2020**, *55*, 8868–8880. [CrossRef]
39. Cortie, M.; Ford, M. A plasmon-induced current loop in gold semi-shells. *Nanotechnology* **2007**, *23*, 235704. [CrossRef]
40. Dasgupta, B.; Ren, Y.; Wong, L.M.; Kong, L.; Tok, E.S.; Chim, W.K.; Chiam, S.Y. Detrimental effects of oxygen vacancies in electrochromic molybdenum oxide. *J. Phys. Chem. C* **2015**, *119*, 10592–10601. [CrossRef]
41. Patel, S.K.S.; Dewangan, K.; Gajbhiye, N.S. Synthesis and room temperature d0 ferromagnetic properties of α-MoO_3 nanofibers. *J. Mater. Sci. Technol.* **2015**, *31*, 453–457. [CrossRef]
42. Lombardi, J.R.; Birke, R.L. A unified approach to surface-enhanced Raman spectroscopy. *J. Phys. Chem. C* **2008**, *112*, 5605–5617. [CrossRef]
43. Fang, Y.; Li, Y.; Xu, H.; Sun, M. Ascertaining p,p′-dimercaptoazobenzene produced from p-aminothiophenol by selective catalytic coupling reaction on silver nanoparticles. *Langmuir* **2010**, *26*, 7737–7746. [CrossRef]
44. Huang, Y.F.; Zhang, M.; Zhao, L.B.; Feng, J.M.; Wu, D.Y.; Ren, B.; Tian, Z.Q. Activation of oxygen on gold and silver nanoparticles assisted by surface plasmon resonances. *Angew. Chem. Int. Ed.* **2014**, *53*, 2353–2357. [CrossRef] [PubMed]
45. Wu, D.Y.; Liu, X.M.; Huang, Y.F.; Ren, B.; Xu, X.; Tian, Z.Q. Surface catalytic coupling reaction of p-Mercaptoaniline linking to silver nanostructures responsible for abnormal SERS enhancement: A DTF study. *J. Phys. Chem. C* **2019**, *113*, 18212–18222. [CrossRef]

nanomaterials

MDPI

Article

A SERS Study of Charge Transfer Process in Au Nanorod–MBA@Cu_2O Assemblies: Effect of Length to Diameter Ratio of Au Nanorods

Lin Guo [1], Zhu Mao [2], Sila Jin [3], Lin Zhu [1], Junqi Zhao [1], Bing Zhao [1,*] and Young Mee Jung [3,*]

1 State Key Laboratory of Supramolecular Structure and Materials, Jilin University, Changchun 130012, China; linguo18@mails.jlu.edu.cn (L.G.); zhulin17@mails.jlu.edu.cn (L.Z.); Zhaojq19@mails.jlu.edu.cn (J.Z.)

2 School of Chemistry and Life Science, Changchun University of Technology, Changchun 130012, China; maozhu@ccut.edu.cn

3 Department of Chemistry, Institute for Molecular Science and Fusion Technology, Kangwon National University, Chuncheon 24341, Korea; jsira@kangwon.ac.kr

* Correspondence: zhaob@mail.jlu.edu.cn (B.Z.); ymjung@kangwon.ac.kr (Y.M.J.)

Abstract: Surface-enhanced Raman scattering (SERS) is a powerful tool in charge transfer (CT) process research. By analyzing the relative intensity of the characteristic bands in the bridging molecules, one can obtain detailed information about the CT between two materials. Herein, we synthesized a series of Au nanorods (NRs) with different length-to-diameter ratios (L/Ds) and used these Au NRs to prepare a series of core–shell structures with the same Cu_2O thicknesses to form Au NR-4-mercaptobenzoic acid (MBA)@Cu_2O core–shell structures. Surface plasmon resonance (SPR) absorption bands were adjusted by tuning the L/Ds of Au NR cores in these assemblies. SERS spectra of the core-shell structure were obtained under 633 and 785 nm laser excitations, and on the basis of the differences in the relative band strengths of these SERS spectra detected with the as-synthesized assemblies, we calculated the CT degree of the core–shell structure. We explored whether the Cu_2O conduction band and valence band position and the SPR absorption band position together affect the CT process in the core–shell structure. In this work, we found that the specific surface area of the Au NRs could influence the CT process in Au NR–MBA@Cu_2O core–shell structures, which has rarely been discussed before.

Keywords: length-to-diameter ratios; core–shell; SERS; surface plasmon resonance; au nanorods; Cu_2O

Citation: Guo, L.; Mao, Z.; Jin, S.; Zhu, L.; Zhao, J.; Zhao, B.; Jung, Y.M. A SERS Study of Charge Transfer Process in Au Nanorod–MBA@Cu_2O Assemblies: Effect of Length to Diameter Ratio of Au Nanorods. *Nanomaterials* **2021**, *11*, 867. https://doi.org/10.3390/nano11040867

Academic Editors: Maurizio Muniz-Miranda and Ronald Birke

Received: 23 February 2021
Accepted: 24 March 2021
Published: 29 March 2021

1. Introduction

Surface-enhanced Raman scattering (SERS) is a rapid, in situ, nondestructive, and ultrasensitive analytical tool. Due to the characteristics of fingerprint recognition, since its discovery in 1974 [1], SERS has been extensively studied [2–6]. As is well known, there are two main mechanisms for SERS enhancement: the electromagnetic mechanism (EM) and the chemical mechanism (CM) [7–9]. For EM mechanism, between noble metal nanoparticles (NPs), surface plasmon resonance (SPR) provides a large electromagnetic field to enable an enhancement factor (EF) up to more than 10^6 [10–12]; for the CM mechanism, the enhancement comes from charge transfer (CT) process between SERS substrates and anchored molecules. Although the EF of the CM is in the range of only 10 to 10^3, it can provide important information about the CT processes between the SERS substrate and molecules that attached on it, which is not provided by the EM mechanism. Due to the nature of the CM mechanism, it can be used for the study of many chemical processes, especially the CT process in metal–semiconductor and semiconductor–semiconductor systems [13–15].

As plasmonic NPs, both gold and silver each have strong localized surface plasmon resonance (LSPR) in the visible region. The SPR of noble metal nanoparticles can be further adjusted by changing the particle shape, size, and surrounding dielectric environment. Among the potential hybrid nanostructures, the metal@semiconductor core–shell heterostructure has been extensively studied because this composite material combines two completely different materials together to form a unique structure with synergistic properties and functions [16,17]. Among these materials, the Au NP@semiconductor structure is widely used in photocatalysis, solar cells, biology, sensing, and other fields [18–22]. Compared with Au NPs, Au nanorods (NRs) are also a commonly used SERS substrate. Au NRs have many advantages, such as two plasmon resonance bands, transverse LSPR and longitudinal LSPR [23,24], and a longitudinal LSPR absorption band that can be flexibly modulated to the required position by changing the experimental conditions. Many researchers have used Au NRs as core–shell nanostructures and semiconductors as shells to adjust the SPR of Au NRs. Au NR cores coupled with semiconductor shells are one of the most widely used structures, but few studies have been applied to SERS.

In this study, a series of Au NRs with different length-to-diameter ratios (L/Ds) were synthesized, and Au NR-4-mercaptobenzoic acid (MBA)@Cu_2O core–shell nanostructures were synthesized. In these core–shell nanostructures, we used the same Cu_2O shell thicknesses to obtain different SPR absorption bands by adjusting the L/Ds of Au NRs. By detecting the SERS spectra of Au NR–MBA structures and Au NR–MBA@Cu_2O structures with 633 and 785 nm laser excitations, we obtained a series of SERS spectra that had information on the CT process in these assemblies. By analyzing the degree of CT (ρ_{CT}), the absorption band of Au NR–MBA@Cu_2O, and the positions of the conduction band (CB) and valence band (VB), we found that the specific surface area of the Au NRs had a certain effect on the CB and VB of the Cu_2O shell in the Au–MBA@Cu_2O structure. The positions of the CB and VB had a significant effect on the CT process in Au NRs and the Cu_2O shell in the core–shell structure.

2. Materials and Methods

2.1. Chemicals

In this study, all chemicals were purchased from Sigma-Aldrich Co., Ltd. (St. Louis, MO, USA and Shanghai, China) in the highest purities available. All chemicals were applied as received without further refinement. Deionized water was used to prepare solutions we employed in this work.

2.2. Sample Preparation

2.2.1. Preparation of Au NRs with Different L/Ds

The series of Au NRs were prepared using a seed-mediated growth method that was well-developed, with some modifications [25]. As a reducing agent, $NaBH_4$ and cetyltrimethylammonium bromide (CTAB)-coated Au seeds were prepared by $HAuCl_4$. CTAB solution (5 mL, 0.1 M) was primarily mixed with $HAuCl_4$ solution (125 μL, 0.01 M), and ice-cold $NaBH_4$ solution (0.3 mL, 0.01 M) was then added via magnetic stirring, making the solution brownish yellow. The Au seeds were prepared at 30 °C for 5 min.

Series volumes of 0.01 M $AgNO_3$ solution (0.05, 0.07, 0.11, 0.15, and 0.21 mL) were added into CTAB solution (20 mL, 0.1 M) at temperature of 30 °C. Then, a $HAuCl_4$ solution (1 mL, 0.01 M) was added to the $AgNO_3$–CTAB solution, and after gentle stirring, ascorbic acid (AA) (0.16 mL, 0.1 M) was added into the mixed solution. The color of the seed growth solution was then gradually changed from dark yellow to colorless. Finally, 48 μL of a solution contained Au NR seeds was added into the growth solution at 30 °C, and the resulting solution was mixed and reacting for 2 h. The prepared Au NRs were centrifuged at 10,000 rpm and redispersed in deionized water, separated twice by repeating the operation.

2.2.2. Preparation of Au NR–MBA with Different L/Ds

The Au NR–MBA solutions with different L/Ds were prepared in one step, adding 1 mL of 10^{-5} M MBA to 1 mL of Au NRs. The solutions were stirred for 5 min. After stirring, the mixtures were centrifuged at 10,000 rpm for 5 min, removing the supernatants.

2.2.3. Preparation of Au NR–MBA@Cu_2O with Different L/Ds

In a typical preparation of Au–MBA@Cu_2O core–shell heterostructures with different L/Ds [26], 0.14 mL of the as-synthesized Au NR–MBA with different L/Ds were added to 5 sample vials, and a sodium dodecyl sulfate (SDS) solution (0.044 g SDS in 4.7 mL of deionized water) was introduced into the vials. Then, 0.05 mL solution of 10^{-3} M $CuCl_2$, 0.125 mL solution of 1 M NaOH, and 0.075 mL solution of $NH_2OH{\cdot}HCl$ were introduced into the reaction systems in the order listed above. NaOH solution was added during growth, the COOH group of MBA was deprotonated, Cu^{2+} and the COO^- group was coordinated, and a crystal nucleus was formed. With the addition of reducing agent ($NH_2OH{\cdot}HCl$), the Cu_2O was grown on the surface of Au NRs, forming a core–shell structure. The solutions turned purple, and finally, as the mixture aged for 2 h, they turned to varying degrees of turquoise. To collect the products and remove the surfactant, we washed all solutions and centrifuged them 2 times with deionized water at 3500 rpm for 5 min.

2.3. *Instruments*

The ultraviolet-visible-near infrared (UV–VIS–NIR) absorption spectra were recorded by a Cary 5000 UV–VIS–NIR spectrometer (Agilent Technologies, Inc., Santa Clara, CA, USA). The surface morphology of the samples was measured by a JEOL JEM-2100 transmission electron microscopy (TEM) system (JEOL Ltd., Tokyo, Japan), operated at an acceleration voltage of 200 kV. The X-ray diffraction (XRD) patterns were obtained by a Siemens D5005 X-ray powder diffractometer (Siemens, Munich, Germany), equipped with a Cu Kα radiation source, operated at 40 kV and 30 mA. In situ Raman spectra were obtained at room temperature by a Jobin Yvon/HORIBA HR evolution Raman spectrometer (HORIBA, Ltd., Koyoto, Japan), equipped with integral BX 41 confocal microscopy. Raman spectra of MBA molecules in the Au NR–MBA@Cu_2O system were accumulated for 30 s and 10 s in the Au NR–MBA system at room temperature. At least 5 Raman measurements were taken for each sample to verify spectral reproducibility. The spectrometer was calibrated by the Raman band of silicon at 520.7 cm^{-1}.

3. Results and Discussion

3.1. *Characterization of Au NRs and Au NR–MBA@Cu_2O with Different L/Ds*

TEM images of Au NRs with different L/Ds are presented in Figure 1a–e. The images clearly show the differences of L/D between these samples and its dependence of the volumes of the added $AgNO_3$ solution.

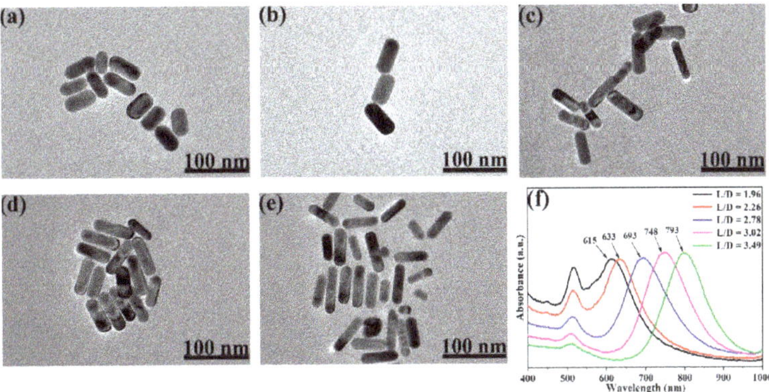

Figure 1. TEM characterizations of different Au nanorods (NRs) synthesized by adding series volumes of $AgNO_3$: (**a**) 0.05 mL, (**b**) 0.07 mL, (**c**) 0.11 mL, (**d**) 0.15 mL, (**e**) 0.21 mL. (**f**) UV–VIS–NIR absorption spectra of Au NRs, synthesized with different volumes of 0.01 M $AgNO_3$ solution (the positions of the maximum absorptions peaks are 615, 633, 693, 748, and 793 nm).

From the TEM images, the average L/Ds of the Au NRs were measured and calculated as 1.96, 2.26, 2.78, 3.02, and 3.49. The size distribution of the Au NRs with different L/Ds is shown in Figures S1 and S2. TEM images of the different L/Ds of Au NR–MBA@Cu_2O core–shell assemblies with the same Cu_2O shells are shown in Figure 2a–e. The distribution of the Cu_2O shell thicknesses is also shown in Figure 2. The XRD pattern for an L/D of 2.78 in the Au NR–MBA@Cu_2O core–shell structure shown in Figure 3 shows that the core-shell structure matched the standard card of cubic Au nanocrystals (JCPDS: 04-0784, space group: $Fm\bar{3}m$, a = 0.4086 nm) and the cubic phase of Cu_2O nanocrystals (JCPDS: 05-0667, space group: $Pn\bar{3}m$, a = 0.4269 nm). The diffraction peaks located at 38.2°, 44.5°, 64.6°, and 77.7° were assigned to the {111}, {200}, {220}, and {311} planes of the face-centered cubic Au nanocrystals. The diffraction peaks located at 2θ = 29.5°, 36.4°, 42.4°, 61.6°, 73.7°, and 77.2° were indexed to the {110}, {111}, {200}, {220}, {311}, and {222} planes of the pure cubic phase of the Cu_2O nanocrystals, respectively. The energy-dispersive X-ray spectroscopy (EDX) spectrum and mapping results of Au NR–MBA@Cu_2O assemblies (shown in Figure S3) suggested a successful assembly of the Au NR–MBA@Cu_2O core–shell structure by the sandwiched MBA molecules. Additionally, no other peaks were observed, indicating a high purity of the combination (Figure S3). On the basis of the TEM images, XRD pattern, and EDX results, we can thus make the conclusion that Cu_2O nanoshells were successfully formed, without additional phases or amorphous structures on the Au NR surfaces.

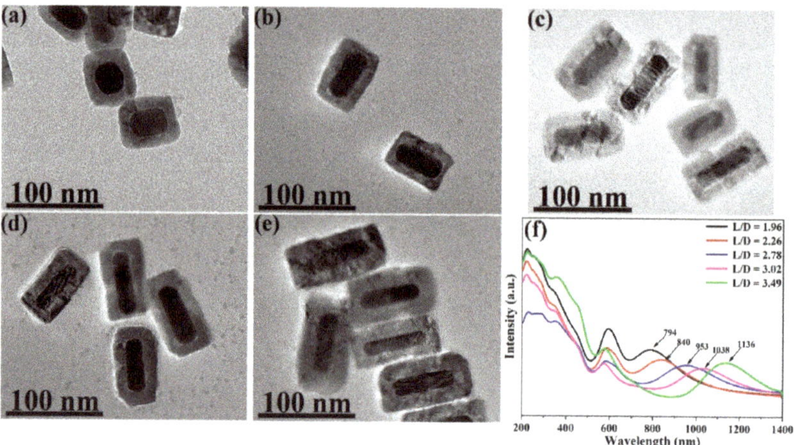

Figure 2. TEM images for different length-to-diameter ratios (L/Ds) ((**a**) 1.96, (**b**) 2.26, (**c**) 2.78, (**d**) 3.02, and (**e**) 3.49) of the Au NR–mercaptobenzoic acid (MBA)@Cu_2O core–shell systems with the same Cu_2O shell thicknesses, and (**f**) UV–VIS–NIR absorption spectra for Au NR–MBA–Cu_2O systems with different L/D grown Cu_2O shells of consistent thickness (the positions of maximum absorption are 794, 840, 953, 1038, and 1136 nm, respectively).

3.2. UV–VIS–NIR Characterization of Au NRs, Au NR–MBA, and Au NR–MBA@Cu_2O Assemblies

UV–VIS–NIR absorption spectra of Au NRs with different L/Ds are shown in Figure 1f. These Au NRs exhibited two LSPR absorption bands of transverse and longitudinal LSPR, and also showed a similar transverse LSPR absorption band at approximately 516 nm and a longitudinal LSPR absorption band that is redshifted with increasing $AgNO_3$ volume in synthesis.

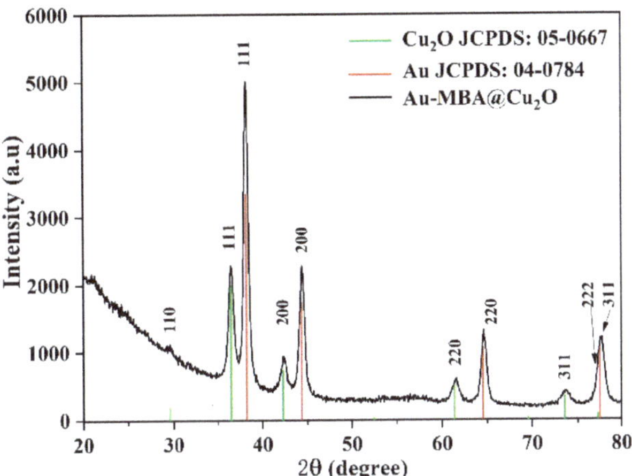

Figure 3. XRD pattern of the Au NR–MBA@Cu_2O (L/D = 2.78) sample.

The UV–VIS–NIR absorption spectra prove that Au NRs with different L/Ds have different plasmon absorption characteristics. The UV–VIS–NIR spectra, shown in Figure 4, present the absorption characteristics of Au NR–MBA assemblies, after the absorption of MBA molecules onto the Au NRs with different L/Ds. After MBA molecules were adsorbed onto the Au NRs by Au–S bonds, compared with Au NRs, the absorption of Au NR–MBA assemblies showed a slightly longer wavelength (the red lines). This comes from the result of dipole–dipole interactions between the Au NRs and MBA molecules, and the changes in the dielectric constant after the Au NRs are coated in MBA molecules. This redshift also suggests that the Au NRs combined successfully with the MBA molecules.

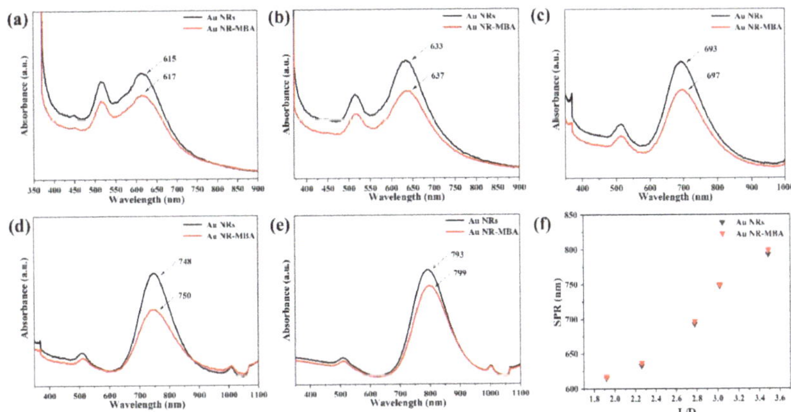

Figure 4. UV–VIS–NIR spectra of the Au NR–MBA assemblies with different Au longitudinal localized surface plasmon resonance (LSPR) absorption bands at (**a**) 617 nm, (**b**) 637 nm, (**c**) 697 nm, (**d**) 750 nm, and (**e**) 799 nm. (**f**) Surface plasmon resonance (SPR) absorption peak position distribution with Au NRs L/D.

After the MBA molecules were attached to the Au NRs, Cu_2O shells were grown on the Au NR–MBA assemblies via the COO^- bonds in the MBA molecules (shown in Figure 2f). Strong absorption was observed at wavelengths shorter than 500 nm, and this absorption was dominated by the interband transition in the Cu_2O shell, which had exciton bands at 223, 287, and 360 nm. Au NRs with different L/Ds grew Cu_2O shells of the same thicknesses; the similar transverse LSPR bands redshifted from 516 nm to 594 nm; and the longitudinal LSPR bands also redshifted from 615, 633, 693, 748, and 793 nm to 794, 840, 953, 1038, and 1136 nm, respectively. The plot of the SPR absorption peak position vs. the L/D of Au NRs is shown in Figure 4f.

3.3. SERS Spectra of MBA in Au NR–MBA and Au NR–MBA@Cu_2O Assemblies

Figure 5c,g shows the SERS spectra of MBA molecules in different Au NR–MBA@Cu_2O assemblies measured with 633 and 785 nm excitations. For better understanding, the SERS spectra of Au NR–MBA are also shown in Figure 5a,e. Table 1 shows the summary of Raman band assignments.

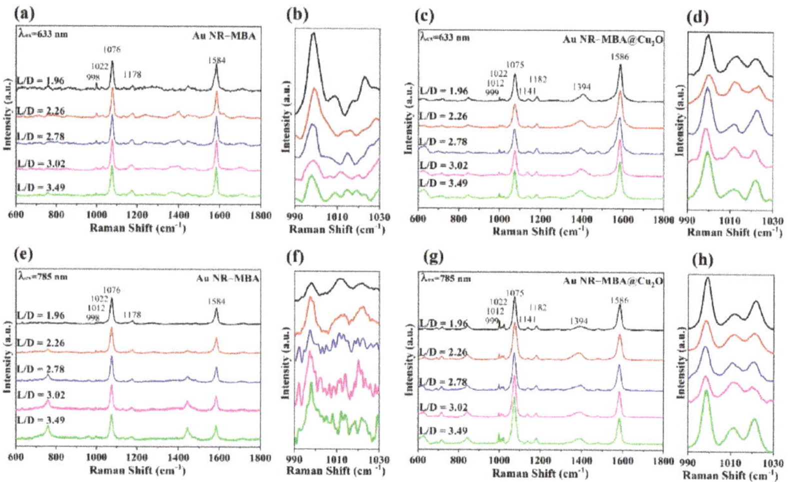

Figure 5. Surface-enhanced Raman scattering (SERS) spectra obtained at (**a**) 633 nm and (**e**) 785 nm laser excitations for the MBA in Au NR–MBA assemblies and SERS spectra obtained at (**c**) 633 nm and (**g**) 785 nm laser excitations for Au NR–MBA@Cu_2O assemblies with the same shell thicknesses. Panels (**b**), (**d**), (**f**), and (**h**) are expanded views of the 990–1030 cm^{-1} regions of (**a**), (**c**), (**e**), and (**g**), respectively.

Table 1. Wavenumbers and bands assignments in the SERS spectrum of the MBA-modified SERS substrate [27,28].

Wavenumber (cm^{-1})		Band Assignments
Au NRs	Assemblies	
998	999	In-plane ring breathing, b_2
1012	1012	In-plane ring breathing + ν(CO), b_2
1022	1022	In-plane ring breathing, b_2
1076	1075	In-plane ring breathing + ν(C–S)
	1141	C–H deformation modes υ_{15}, b_2
1178	1182	C–H deformation modes υ_9, a_1
	1394	$\nu(COO^-)$
1584	1586	Totally symmetric ν(CC), a_1
1710		C=O stretching

ν, stretching. For ring vibrations, the corresponding vibrational modes of benzene and the symmetry species with C_{2v} symmetry are indicated.

For the MBA molecules in the Au NR–MBA assemblies, two important bands attributed to a_1-type vibrations are observed. One band at 1076 cm^{-1} was assigned to the in-plane ring breathing mode coupled with ν(C-S), and another band at 1584 cm^{-1} was ascribed to the totally symmetric ν(CC) mode. Bands at 998, 1012, and 1022 cm^{-1} were assigned to in-plane ring breathing, assignment to b_2-type vibrations. After the coating shell of Cu_2O was coated on the Au NR–MBA system, spectral changes were observed in these bands. These shifts are clearly described in Figure 5 and Table 1.

It can be observed that the band at 1394 cm^{-1} (classified as the COO^- stretching mode) increased after the Cu_2O shell was introduced in the system. This was because after the introduction of Cu_2O shell, the original COO^- stretching mode had been promoted. At the same time, the intensity of the band at 1710 cm^{-1} (caused by the C=O stretching mode) decreased after the introduction of the Cu_2O shell. All Raman spectra were normalized to the band at 1074 cm^{-1} in Figure 5 in order to facilitate a direct comparison. Interestingly, some changes were observed in the relative intensities of the b_2-type vibration bands in Au NR–MBA and Au NR–MBA@Cu_2O. Figure 5b,d,f,h show the SERS spectra in the 990–1030 cm^{-1} region to detect the relative intensity changes more clearly.

Lombardi et al. developed a CT model of SERS chemical mechanism [29,30], in which the Franck–Condon contribution can only enhance the full symmetric vibration modes of the probe molecules, while the Herzberg–Teller effect can enhance the totally symmetric and nontotally symmetric vibration modes. The CT contribution of the system is usually determined by the ratio of the non-totally symmetric vibration modes to the totally symmetric vibration modes. Lombardi and Birke defined the ρ_{CT} for each mode as a quantitative calculation of relative CT contribution to the intensity of SERS as

$$\rho_{CT}(k) = \frac{I^k(CT) - I^k(SPR)}{I^k(CT) + I^0(SPR)}$$

where k is used as an index to determine the individual molecular lines in the Raman spectrum. Two intensities of reference lines in the spectral region without CT contributions is needed in the equation. One intensity is I^k(SPR), that is, the intensity of the line (k) under consideration, in which only SPR contributes to the SERS intensity. The other reference is a totally symmetrical line, which is also measured by the SPR contribution, and denoted as I^0(SPR). I^k(CT) is the line intensity (k) measured in the spectral region, where CT resonance contributes extra to the SERS intensity.

In this study, the bands at 999 (in-plane ring breathing, b_2) and 1182 cm^{-1} (C–H deformation modes v_9, a_1) in the Au NR–MBA@Cu_2O system (corresponding to the bands at 998 and 1178 cm^{-1} in the Au NR–MBA system) were selected to compare the CT contributions. Different laser excitations were employed for testing the CT process.

In Figure 6a,c, the plots demonstrate the trend of the I_{998}/I_{1178} ratio and ρ_{CT} in Au NR–MBA assemblies under 633 and 785 nm excitations. For the Au NR–MBA system, the ratio of I_{998}/I_{1178} and ρ_{CT} tended to decrease with the increasing L/D of Au NRs under 633 nm laser excitation. Compared with Figure 4, with the increasing L/D of Au NRs, the absorption bands of the Au NR–MBA assembly deviated from 633 nm, which means that the absorption bands no longer matched the incident laser. This mismatching decreased the resonance of Au NRs and the incident laser, leading to a decrease in the CT from Au NRs to MBA molecules. At the same time, the ratio of I_{998}/I_{1178} and ρ_{CT} simultaneously decreased. In Figure 6c, the ratio of I_{998}/I_{1178} and ρ_{CT} increased with the increasing L/D of Au NRs. The redshift process in the absorption bands was closer to the 785 nm laser line, resulting in a stronger resonance of Au NRs and the incident laser. Intense coupling of Au NRs and the incident laser enabled prominent CT from the Au NRs to the MBA molecules, and the ratio of I_{998}/I_{1178} and ρ_{CT} at 785 nm laser excitation showed an increasing trend. Figure 6b,d presents the trends of the ratio of I_{999}/I_{1182} and ρ_{CT} in Au NR–MBA@Cu_2O assemblies. In the absorption bands of Au NR–MBA@Cu_2O assemblies shown in Figure 2f, the longitudinal LSPR absorption bands redshifted from 794 to 1136 nm, and the transverse LSPR was sustained at approximately 600 nm. As shown in Figure 6b, the ratio of I_{999}/I_{1182} and ρ_{CT} showed a random trend at 633 nm laser excitation. In this region, the longitudinal SPR absorption bands of the assemblies far away from the laser line made a very small contribution to the SERS signal and the CT process. At the same time, the changes in transverse LSPR were not obvious and had little effect on the CT process. It is worth noting that the ratio of I_{998}/I_{1178} and ρ_{CT} of the last sample (L/D = 3.49) increased slightly. We calculated the specific surface area of Au NRs (as shown in Table S1) and we can see that the specific surface area of Au NRs had a tendency to increase, especially in the last sample, where the specific surface area increased sharply. This increase in the specific surface area meant an increase in the surface state, which may have been the reason for the increase in the ratio of I_{998}/I_{1178} and ρ_{CT} for the last sample. For 785 nm laser excitation, Figure 6d shows a tendency to decrease, except for samples with an L/D of 3.49. To determine the reason for this decrease, we measured the ultraviolet photoelectron spectroscopy (UPS) of these samples.

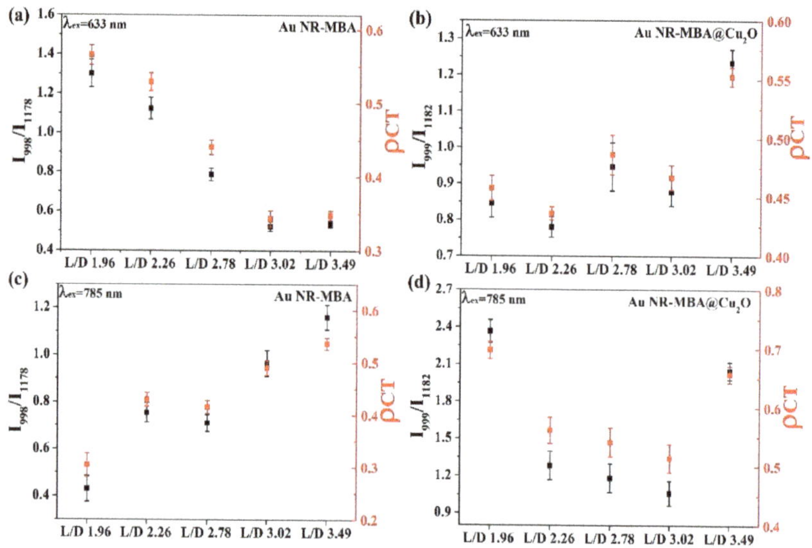

Figure 6. Degree of charge transfer (CT) (ρ_{CT}) in the Au NR–MBA and Au NR–MBA@Cu_2O samples and the SERS intensity ratio between the bands at 998 (b_2) and 1178 cm^{-1} (a_1) (999 (b_2), and 1182 cm^{-1} (a_1) of the Au NR–MBA (Au NR–MBA@Cu_2O) assemblies at two laser excitations: (**a**,**b**) 633 nm and (**c**,**d**) 785 nm.

3.4. UPS of Au NR–MBA@Cu_2O Assemblies with Different L/Ds

According to the UPS results, the Fermi level of the Au NRs with L/Ds of 1.96, 2.26, 2.78, 3.02, and 3.49 corresponded to 3.94, 3.99, 4.01, 4.03, and 4.10 eV from the vacuum level, respectively (shown in Figure S4). The lowest unoccupied molecular orbital (LUMO) for MBA was located at 2.99, and the highest occupied molecular orbital (HOMO) levels were located at 7.65 eV (shown in Figures S5 and S6). The CB and VB levels of pure Cu_2O were 5.7 and 7.9 eV, respectively (shown in Figure S7). We also used UPS to investigate Au NR–MBA@Cu_2O assemblies with different L/Ds. Figures S8–S12 and Table 2 show the UPS results and the energy levels of CB and VB values, respectively.

Table 2. The energy level of pure Cu_2O and Au NR–MBA@Cu_2O assemblies with different L/Ds.

L/D	1.96	2.26	2.78	3.02	3.49	Pure Cu_2O
CB (eV)	−5.56	−5.76	−6.04	−6.06	−5.86	−5.7
VB (eV)	−7.76	−7.96	−8.24	−8.26	−8.06	−7.9

The LUMO of MBA molecules (−2.99 eV) was much higher than the VB of Cu_2O (−5.56 eV, for the highest one of the samples), which made the trend that charge transfers to the Cu_2O from MBA molecules. In Figure 6, for 633 nm laser line (Figure 6a,b), compared with Au NR-MBA assembly, the CT degrees of samples 3, 4, and 5 (L/D 2.78, 3.02, and 3.49) in Au NR–MBA@Cu_2O core–shell structure were significantly increased. Since LSPR absorption bands of Au NR were far away from 633 nm, the resonance effect of Au NR and laser line can be ignored, and thus the reason for the elevation of ρ_{CT} should be the CT from MBA molecule to the Cu_2O. It is abnormal for samples 1 and 2 (L/D 1.96 and 2.26). In these two samples, the absorption bands of Au NR–MBA assembly were too close to the 633 nm laser line, producing a strong resonance between Au NR–MBA assemblies and the laser line, resulting in a large ρ_{CT}. However, in Au NR–MBA@Cu_2O assembly, LSPR absorption bands were far away from 633 nm laser line, and thus the contribution of resonance between Au NR–MBA@Cu_2O assembly and laser line can be ignored. Therefore, samples 1 and 2 (L/D 1.96 and 2.26, respectively) were abnormal in all these samples. The same phenomenon for the same samples in laser line of 785 nm can be seen in Figure 6c,d. This was the main reason that we confirmed that the CT took place from the Au NR to MBA to Cu_2O. In this system, according to previous studies, the CT from Au NRs to Cu_2O shell happens in about several picoseconds, and the system becomes equilibrated. This is a dynamic equilibrium, which means the CT process occurs all the time during the excitation of laser, and flowing of electrons make the system equilibrium.

As shown in Table 2, the CB and VB values exhibited a tendency to decrease in the first four samples, but in the last sample, the CB and VB values increased suddenly, which made the VB value close to the HOMO value of MBA molecules. Theoretically, this tendency should lead to a Au NR–MBA@Cu_2O ρ_{CT} value between the second sample (L/D = 2.26) and the third sample (L/D = 2.78). However, there was another factor that affected the final ρ_{CT} value, this being the specific surface area of Au NRs shown in Table S1. In the last sample (L/D = 3.49), the length of Au NRs did not change significantly with increasing L/D, and in contrast, the radius of Au NRs decreased, resulting in a noticeable increase in the specific surface area of Au NRs. As the specific surface area of Au NRs increased, the volume of Au NRs decreased, which led to an increase in the CB and VB values for Cu_2O in Au NR–MBA@Cu_2O, making this value close to that of pure Cu_2O. Returning to the ρ_{CT} value, as the specific surface area of Au NRs increased, the surface state showed a nonnegligible facilitation of the CT process, which significantly increased the ρ_{CT} value.

Figure 7 presents the numerical relationship between the Fermi level of Au NRs, LUMO and HOMO of the MBA molecules, and CB and VB values of the Cu_2O shell. Arrows indicate the direction of electron transfer in the assemblies, which occurred from the Au NRs to the LUMO of the MBA molecules, and then to the CB of the Cu_2O shell. The energy gap between the LUMO of the molecules and the CB of Cu_2O significantly influenced the CT process. The larger the energy gap was, the more difficult the CT process. The trend of the CB value of Cu_2O perfectly matched that of the ratio of I_{999}/I_{1182} and ρ_{CT}. The results clearly showed that the location of the CB of Cu_2O can affect the CT process as well as the SPR position.

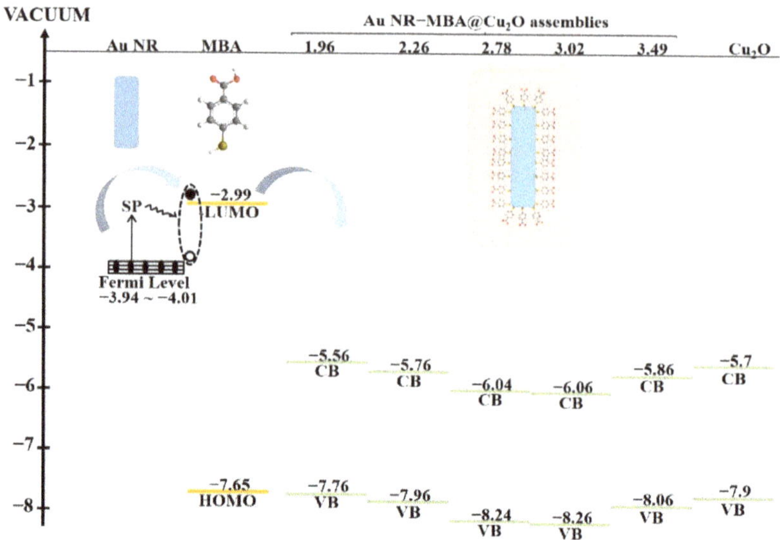

Figure 7. Energy level of Au NRs, MBA, Cu_2O, and Au NR–MBA@Cu_2O with different L/Ds.

4. Conclusions

In this work, we successfully synthesized a series of Au NR–MBA@Cu_2O complexes with different L/Ds, which each had a consistent Cu_2O shell thickness of approximately 15 nm. We adjusted the SPR absorption band by changing the L/Ds of the core Au NRs. In these assemblies, charge can be transferred between the Au NRs and the Cu_2O shell, with the MBA molecule acting as a bridge. To study the effect of the SPR absorption band on the CT process in metal and semiconductor core–shell structures, we measured the SERS spectra of these assemblies with 633 and 785 nm laser excitations. On the basis of an analysis of the SERS spectra of these assemblies, we can conclude that coupling of the SPR absorption band and the incident laser light had a significant effect on the CT process. The degree of CT increased with coupling of the incident laser light and the SPR absorption band. At the same time, we demonstrated that the specific surface area of Au NRs is another important factor influencing the CT process when the SPR absorption band is approximately coupled to the incident laser light. The greater the specific surface area of Au NRs, the greater the degree of CT. These results will provide a new guidance for further exploration of the CT process between metal nanoparticles and semiconductors.

Supplementary Materials: The following are available online at https://www.mdpi.com/article/10.3390/nano11040867/s1, Figure S1: Size distribution of the Au NRs with different L/Ds. Figure S2: Size distribution of the Cu_2O shell thicknesses with different L/Ds. Figure S3: Elemental mapping and energy-dispersive X-ray spectroscopy (EDX) spectrum of Au NR–MBA@Cu_2O assemblies. Figure S4: The UPS spectra of Au NRs with different L/Ds. Figure S5: The UPS spectra of MBA. Figure S6: The UV–VIS spectrum of MBA. Figure S7: The UPS spectra of Cu_2O. Figure S8–S12: The UPS spectra of Au NR–MBA@Cu_2O with different L/Ds. Table S1: Specific surface area statistics for Au NR with different L/Ds.

Author Contributions: Methodology, L.G. and Z.M.; data analysis, L.G., S.J., L.Z., and J.Z.; writing—original draft preparation, L.G. and Z.M.; writing—review and editing, B.Z. and Y.M.J.; project administration, B.Z. and Y.M.J.; funding, B.Z. and Y.M.J. All authors have read and agreed to the published version of the manuscript.

Funding: The research was supported by the National Natural Science Foundation (grant nos. 21773080 and 21711540292) of the People's China. This work was also supported by the National Research Foundation of Korea (NRF) grant funded by the Korean government (nos. NRF-

2018R1A2A3074587, NRF-2020R1A4A1016093, and NRF-2020K2A9A2A06036299) and the Korea Basic Science Institute (National Research Facilities and Equipment Center) grant funded by the Ministry of Education (no. 2020R1A6C101A195). Raman spectra were measured in Kangwon Radiation Convergence Research Support Center of Korea Basic Science Institute (KBSI) at Kangwon National University.

Institutional Review Board Statement: Not applicable.

Informed Consent Statement: Not applicable.

Data Availability Statement: The data presented in this study are available from the corresponding author.

Conflicts of Interest: The authors declare no conflict of interest.

References

1. Fleischmann, M.; Hendra, P.J.; McQuillan, A.J. Raman spectra of pyridine adsorbed at a silver electrode. *Chem. Phys. Lett.* **1974**, *26*, 163–166. [CrossRef]
2. Shen, Z.D.; Wang, H.Y.; Yu, Q.; Li, Q.; Lu, X.M.; Kong, X.M. On-site separation and identification of polycyclic aromatic hydrocarbons from edible oil by TLC-SERS on diatomite photonic biosilica plate. *Microchem. J.* **2021**, *160*, 105672. [CrossRef]
3. Ngo, H.T.; Wang, H.-N.; Fales, A.M.; Vo-Dinh, T. Label-Free DNA Biosensor Based on SERS Molecular Sentinel on Nanowave Chip. *Anal. Chem.* **2013**, *85*, 6378–6383. [CrossRef]
4. Nie, X.M.; Chen, Z.Y.; Tian, Y.P.; Chen, S.; Qu, L.L.; Fan, M.B. Rapid detection of trace formaldehyde in food based on surface-enhanced Raman scattering coupled with assembled purge trap. *Food Chem.* **2021**, *340*, 127930. [CrossRef]
5. Chen, Q.; Shi, C.L.; Qin, L.X.; Kang, S.-Z.; Li, X.Q. A low-cost 3D core-shell nanocomposite as ultrasensitive and stable surface enhanced Raman spectroscopy substrate. *Sens. Actuators B* **2021**, *327*, 128907. [CrossRef]
6. Bazzaoui, E.A.; Aubard, J.; Félidj, N.; Laurent, G.; Lévi, G. Ex situ and in situ SERS analyses of polybithiophene using roughened Ag and Cu electrodes and multilayer SERS-active systems. *J. Raman Spectrosc.* **2005**, *36*, 817–823. [CrossRef]
7. Campion, A.; Kambhampati, P. Surface-enhanced Raman scattering. *Chem. Soc. Rev.* **1998**, *27*, 241–250. [CrossRef]
8. Otto, A. The 'chemical' (electronic) contribution to surface-enhanced Raman scattering. *J. Raman Spectrosc.* **2005**, *36*, 497–509. [CrossRef]
9. Das, R.S.; Agrawal, Y.K. Raman spectroscopy: Recent advancements, techniques and applications. *Vib. Spectrosc.* **2011**, *57*, 163–176. [CrossRef]
10. Wustholz, K.L.; Henry, A.-I.; McMahon, J.M.; Freeman, R.G.; Valley, N.; Piotti, M.E.; Natan, M.J.; Schatz, G.C.; Duyne, R.P.V. Structure-Activity Relationships in Gold Nanoparticle Dimers and Trimers for Surface-Enhanced Raman Spectroscopy. *J. Am. Chem. Soc.* **2010**, *132*, 10903–10910. [CrossRef]
11. Prinz, J.; Heck, C.; Ellerik, L.; Merk, V.; Bald, I. DNA origami based Au-Ag-core-shell nanoparticle dimers with single-molecule SERS sensitivity. *Nanoscale* **2016**, *8*, 5612–5620. [CrossRef]
12. Tran, T.T.; Herfort, D.; Jakobsen, H.J.; Skibsted, J. Site Preferences of Fluoride Guest Ions in the Calcium Silicate Phases of Portland Cement from Si-29{F-19} CP-REDOR NMR Spectroscopy. *J. Am. Chem. Soc.* **2009**, *131*, 14170. [CrossRef]
13. Wang, Y.; Liu, J.; Ozaki, Y.; Xu, Z.R.; Zhao, B. Effect of TiO_2 on Altering Direction of Interfacial Charge Transfer in a TiO_2-Ag-Mpy-FePc System by SERS. *Angew. Chem. Int. Ed.* **2019**, *58*, 8172–8176. [CrossRef] [PubMed]
14. Li, P.; Zhu, L.; Ma, C.; Zhang, L.X.; Guo, L.; Liu, Y.W.; Ma, H.; Zhao, B. Plasmonic Molybdenum Tungsten Oxide Hybrid with Surface-Enhanced Raman Scattering Comparable to that of Noble Metals. *ACS Appl. Mater. Interfaces* **2020**, *12*, 19153–19160. [CrossRef] [PubMed]
15. Ji, W.; Li, L.F.; Song, W.; Wang, X.N.; Zhao, B.; Ozaki, Y. Enhanced Raman Scattering by ZnO Superstructures: Synergistic Effect of Charge Transfer and Mie Resonances. *Angew. Chem. Int. Ed.* **2019**, *58*, 14452–14456. [CrossRef]
16. Liu, Y.W.; Ma, H.; Han, X.X.; Zhao, B. Metal-semiconductor heterostructures for surface-enhanced Raman scattering: Synergistic contribution of plasmons and charge transfer. *Mater. Horiz.* **2021**, *8*, 370–382. [CrossRef]
17. Gao, W.Q.; Liu, Q.; Zhang, S.; Yang, Y.Y.; Zhang, X.F.; Zhao, H.; Qin, W.; Zhou, W.J.; Wang, X.N.; Liu, H.; et al. Electromagnetic induction derived micro-electric potential in metal-semiconductor core-shell hybrid nanostructure enhancing charge separation for high performance photocatalysis. *Nano Energy* **2020**, *71*, 104624. [CrossRef]
18. Wan, X.D.; Liu, J.; Wang, D.; Li, Y.M.; Wang, H.Z.; Pan, R.G.; Zhang, E.H.; Zhang, X.M.; Li, X.Y.; Zhang, J.T. From core-shell to yolk-shell: Keeping the intimately contacted interface for plasmonic metal@semiconductor nanorods toward enhanced near-infrared photoelectrochemical performance. *Nano Res.* **2020**, *13*, 1162–1170. [CrossRef]
19. Lin, H.Y.; Chen, Y.F.; Wu, J.G.; Wang, D.I.; Chen, C.C. Carrier transfer induced photoluminescence change in metal-semiconductor core-shell nanostructures. *Appl. Phys. Lett.* **2006**, *88*, 161911. [CrossRef]
20. Guo, H.X.; Su, X.P.; Su, Q.F.; Zhuang, W.; You, Z.J. Au-coated Fe_3O_4 core-shell nanohybrids with photothermal activity for point-of-care immunoassay for lipoprotein-associated phospholipase A_2 on a digital near-infrared thermometer. *Anal. Bioanal. Chem.* **2021**, *413*, 235–244. [CrossRef] [PubMed]

21. Lv, Y.P.; Duan, S.B.; Zhu, Y.C.; Yin, P.; Wang, R.M. Enhanced OER Performances of Au@$NiCo_2S_4$ Core-Shell Heterostructure. *Nanomaterials* **2020**, *10*, 611. [CrossRef]
22. Zhu, S.L.; Deng, D.; Nguyen, M.T.; Rachel Chau, Y.-T.; Wen, C.-Y.; Yonezawa, T. Synthesis of Au@Cu_2O Core-Shell Nanoparticles with Tunable Shell Thickness and Their Degradation Mechanism in Aqueous Solutions. *Langmuir* **2020**, *36*, 3386–3392. [CrossRef] [PubMed]
23. Yu, Y.-Y.; Chang, S.-S.; Lee, C.-L.; Chris Wang, C.R. Gold Nanorods: Electrochemical Synthesis and Optical Properties. *J. Phys. Chem. B* **1997**, *101*, 6661–6664. [CrossRef]
24. Huang, X.H.; Neretina, S.; El-Sayed, M.A. Gold nanorods: From synthesis and properties to biological and biomedical applications. *Adv. Mater.* **2009**, *21*, 4880–4910. [CrossRef]
25. Nikoobakht, B.; El-Sayed, M.A. Preparation and Growth Mechanism of Gold Nanorods (NRs) Using Seed-Mediated Growth Method. *Chem. Mater.* **2003**, *15*, 1957–1962. [CrossRef]
26. Kuo, C.-H.; Hua, T.-E.; Huang, M.H. Au Nanocrystal-Directed Growth of Au-Cu_2O Core-Shell Heterostructures with Precise Morphological Control. *J. Am. Chem. Soc.* **2009**, *131*, 17871–17878. [CrossRef]
27. Li, R.; Lv, H.; Zhang, X.; Liu, P.; Chen, L.; Cheng, J.; Zhao, B. Vibrational spectroscopy and density functional theory study of 4-mercaptobenzoic acid. *Spectrochim. Acta Part A* **2015**, *148*, 369–374. [CrossRef] [PubMed]
28. Zhang, X.; Yu, Z.; Ji, W.; Sui, H.M.; Cong, Q.; Wang, X.; Zhao, B. Charge-Transfer Effect on Surface-Enhanced Raman Scattering (SERS) in an Ordered Ag NPs/4-Mercaptobenzoic Acid/TiO_2 System. *J. Phys. Chem. C* **2015**, *119*, 22439–22444. [CrossRef]
29. Lombardi, J.R.; Birke, R.L. A Unified Approach to Surface-Enhanced Raman Spectroscopy. *J. Phys. Chem. C* **2008**, *112*, 5605–5617. [CrossRef]
30. Lombardi, J.R.; Birke, R.L. A Unified View of Surface-Enhanced Raman Scattering. *Acc. Chem. Res.* **2009**, *42*, 734–742. [CrossRef]

nanomaterials

MDPI

Article

Investigations of Shape, Material and Excitation Wavelength Effects on Field Enhancement in SERS Advanced Tips

Yaakov Mandelbaum [1], Raz Mottes [1], Zeev Zalevsky [2,3], David Zitoun [4] and Avi Karsenty [1,5,*]

1 Advanced Laboratory of Electro-Optics (ALEO), Department of Applied Physics/Electro-Optics Engineering, Jerusalem College of Technology, Jerusalem 9116001, Israel; ymandelb@g.jct.ac.il (Y.M.); rm1995ex@gmail.com (R.M.)
2 Faculty of Engineering, Bar-Ilan University, Ramat Gan 5290002, Israel; Zeev.Zalevsky@biu.ac.il
3 Nanotechnology Center, Bar-Ilan University, Ramat Gan 5290002, Israel
4 Faculty of Exact Science, Department of Chemistry, Bar-Ilan University, Ramat Gan 5290002, Israel; David.Zitoun@biu.ac.il
5 Nanotechnology Center for Research and Education, Jerusalem College of Technology, Jerusalem 9116001, Israel
* Correspondence: karsenty@jct.ac.il; Tel.: +972-2-675-1140

Abstract: This article, a part of the larger research project of Surface-Enhanced Raman Scattering (SERS), describes an advanced study focusing on the shapes and materials of Tip-Enhanced Raman Scattering (TERS) designated to serve as part of a novel imager device. The initial aim was to define the optimal shape of the "probe": tip or cavity, round or sharp. The investigations focused on the effect of shape (hemi-sphere, hemispheroid, ellipsoidal cavity, ellipsoidal rod, nano-cone), and the effect of material (Ag, Au, Al) on enhancement, as well as the effect of excitation wavelengths on the electric field. Complementary results were collected: numerical simulations consolidated with analytical models, based on solid assumptions. Preliminary experimental results of fabrication and structural characterization are also presented. Thorough analyses were performed around critical parameters, such as the plasmonic metal—Silver, Aluminium or Gold—using Rakic model, the tip geometry—sphere, spheroid, ellipsoid, nano-cone, nano-shell, rod, cavity—and the geometry of the plasmonic array: cross-talk in multiple nanostructures. These combined outcomes result in an optimized TERS design for a large number of applications.

Keywords: TERS; SERS; nano-cones; nano-cavities; plasmon; numerical; analytical model

Citation: Mandelbaum, Y.; Mottes, R.; Zalevsky, Z.; Zitoun, D.; Karsenty, A. Investigations of Shape, Material and Excitation Wavelength Effects on Field Enhancement in SERS Advanced Tips. *Nanomaterials* **2021**, *11*, 237. https://doi.org/10.3390/nano11010237

Received: 5 January 2021
Accepted: 14 January 2021
Published: 18 January 2021

Publisher's Note: MDPI stays neutral with regard to jurisdictional claims in published maps and institutional affiliations.

1. Introduction

The need for the development of real-time sensors, capable of monitoring continuous flow reactions and phenomena, became one of the next challenging frontiers to reach in chemical sensing. This is why imaging sensors, capable of recording and reporting spatial variations in real time, are more than desirable. The Surface Enhanced Raman Scattering (SERS) method, capable of chemical sensing, is used either on chemicals which are adsorbed on a particular substrate, by scanning with a sharp metallic tip [1,2], or by dispersing metallic nano-particles into the solution [3]. The first and second methods preclude real time detection. The first method is not capable of self-refreshing, and the second method due to lengthy scanning times; the third method is not position specific. Thus, the development of an imaging sensor which is both dynamic and has specificity in space is more than justifiable. It is here envisioned as an array of SERS nanostructures, tips or cavities, with the capability of fulfilling these demands.

Fleischmann et al. first observed SERS from pyridine adsorbed on electrochemically roughened silver in 1973 [4]. A few years later, several research teams working in parallel, arrived at the same observations, noting that the scattering species concentration could not explain the enhanced signal. In fact, each team suggested a different approach to explain the observed phenomenon, and the explanatory mechanisms they proposed for the

SERS effect are still accepted today. Jeanmaire and Van Duyne [5] proposed a theoretical mechanism by which Raman signals are amplified by an electric field enhancement near a metallic surface.

When a substrate is impinged by incident light source, the phenomenon generates an excitation of the localized surface plasmons. The electric field enhancement produced near the surface, is maximized in the resonant condition when the frequency of the incident light is equal to the surface plasmon frequency. Thus, Raman signals' intensity for the adsorbates is increased due to electric field enhancement near the substrate. The size, shape and material of the nano-particles determine the electromagnetic enhancement of the SERS, which is theoretically calculated in order to enable factor values in the range of $\sim 10^{10}$–10^{11} [6]. This enhancement factor [EF] can be approximated by the magnitude of the localized electromagnetic field to the fourth power (the E^4 approximation approach will be further discussed in this article).

While Jeanmaire and Van Duyne [5] proposed an explanation based on an electromagnetic effect, Albrecht and Creighton [7] proposed an alternative explanation based on charge-transfer effect. Raman spectrum peaks are enhanced due to intermolecular and intramolecular charge transfers. The high-intensity charge transfers from the metal surface with wide band to the adsorbing species causing great enhancement for species adsorbing the metal surface [8]. Because surface plasmon appears only in metal surface, with near-zero band gaps, the effect of Raman resonance enhancement is dominant in SERS for species on small nanoclusters with considerable band gaps. Although the charge-transfer effect explanation is less accepted than the electromagnetic effect, both effects can probably occur together for metal surfaces [9]. At the end, Ritchie predicted the surface plasmon's existence years before [10].

The methods to perform and prepare SERS measurements have progressed with time, moving from electrochemically roughened silver [9], distribution of metal nano-particles on the surface [11], lithography [12], porous silicon support [13,14], and two-dimensional silicon nano-pillars embedded with silver [15]. The most common method consists today of liquid sample deposition onto a silicon or glass surface, with a nanostructured noble metal surface. The enhancement phenomenon is intensely affected by the geometry (both shape and size) of the metal nano-particles, since the ratio of absorption and scattering events are influenced by these factors [16,17]. According to Bao et al., it may be an ideal particles size and an ideal surface thickness as a function of each particular experiment [18]. For example, while the excitation of non-radiative multipoles is a result of exceptionally large particles, the loss of electrical conductance, and as a consequence the lack of field enhancement, is due to extremely small particles. For particles sharing only the size of a few atoms, there must be a large collection of electrons to oscillate together, since there is no defined plasmon [19]. Higher-order transitions cause the enhancement's overall decrease in efficiency, since the dipole transition leads to Raman scattering. Both high uniformity and field enhancement define ideal SERS substrates, which are fabricated on wafer scale with label-free super resolution microscopy. Such a resolution was proven adequate, when using SERS signal's fluctuations on such uniform and high-performance plasmonic meta-surfaces [20].

With the large development of SERS usage, and since a vast amount of literature became available, several publications focused in the last two decades on reviewing specific SERS sub-domains. While Pamela Mosier-Boss reviewed the substrates for chemical sensing [21], additional summarizing studies focused on substrates and analytes [22]. Others focused also on Surface-Enhanced Resonance Raman Scattering (SERRS), and on possible applications [23]. Van Duyne and other pioneers in the field presented a large perspective on the present and future achievements in SERS, both in the past [24] and more recently [25]. However, large-scale and methodical analysis of Tip-Enhanced Raman Spectroscopy (TERS) has not yet presented, while combining numerical and analytical complementary analyses. This is why this current study largely focuses, among others, on the investigations of TERS effects of shape, material and excitation wavelength on field enhancement (FE).

As part of the Surface-Enhanced Raman Spectroscopy (SERS) technique, one can find a more accurate approach entitled Tip-Enhanced Raman Spectroscopy (TERS). TERS is the combination of a scanning probe microscope and a plasmonic metal tip, and its strongest point, aside from its high chemical sensitivity, is the high spatial resolution (beyond the diffraction limit), and imaging it can provide information for data analysis. In this technique, Raman scattering enhancement occurs only at the extremity of a near atomically sharp pin, usually coated with gold [26]. In SERS spectroscopy, there are two limitations:

1. The signal is produced by the sum of a large number of molecules.
2. The resolution is limited to Abbe limit, which is half the wavelength of the incident light.

TERS overcomes these limitations by sampling only a small number of molecules near the tip, which consists of a few tens of nanometers. There are basically two kinds of TERS, which are generally accepted by the TERS community:

1. The aperture type—using a fiber whose hollow core acts as an aperture for the light;
2. The apertureless type—that uses a sharp tip. Near-field scanning optical microscopy (NSOM) is the general term for STM, AFM and even SFM-TERS. While ANSOM (Aperture NSOM) is mainly for fiber-type tips or cantilever tips with a hole at the tip end.

Sometimes, it also results that TERS is combined with other methods:

1. TERS can be used in scanning probe microscopy (SPM).
2. TERS can also be coupled to a scanning tunneling microscope (STM-TERS). In such a case, the enhancement will be produced by the gap mode plasmon between the metallic probe and the metallic substrate [27].
3. Raman microscope coupled with atomic force microscope (AFM-TERS), which is widely used in live bio samples [28].
4. Shear force microscopy based TERS system (SFM-TERS).
5. Near-field scanning optical microscopy (NSOM) based TERS system (NSOM-TERS). Raman signals are collected through the same fiber that delivers the excitation light.

TERS has been used for several applications: imaging of single atoms, imaging of internal molecular structure [29–32], imaging of vibrational normal modes of single porphyrin molecules [33], demonstration of DNA sequencing [34], and ion-selective, atom-resolved imaging of a 2D Cu_2N insulator using a functionalized tip [35]. Several works looked at the geometry of the tips for specific optical resonance and enhancement purposes [36] and for large opening angles [37].

In this research, the main goal is to optimize the tip nanostructure geometry and material, towards obtaining the optimal density of multiple nanostructures per future array. The final planned pixel will be built with a nanostructured substrate composed of an array of projections or cavities. The shape of these nanostructures and the thickness of their metallic layer (Ag, Au, and Al) can be tuned to deliver the maximal enhancement at the desired wavelength. The number and arrangement of nanostructures was optimized to obtain maximal responsivity.

2. Numerical Method: The Finite Element Method (FEM) for PDEs

2.1. Best Known Methods (BKM) Choice and Usage

Complementary methods, analytical and numerical, were used in order to accurately model SERS. In this research, the primary numerical approach was the Finite Elements Method (FEM), applied in COMSOL platform tool, and combined with algorithmic optimization algorithms such as Simulated Annealing and Method of Simplexes. Additional simulation programs such as CST, DDSCAT, and MEEPS, based on alternate methods like Method of Moments (MoM), Discrete Dipole Approximation (DDA) and Finite Differences Time Domain (FDTD), respectively, have been considered as necessary, but are not presented in this study. The enhancement of Raman emission from emitters, which are volume-dispersed in a fluid, as well as the possibility of near-field detection through plasmonic antennae, will require the use of simulation approaches, which go beyond the current approach based on surface-integrals in the E^4 approximation. Numerical simulation of propagation of incoherent radiation are performed using a Monte-Carlo approach for individual source phases, as well as a continuum model.

2.2. Mesh Shapes and Sizes

Finite Elements Method (FEM) is used in multi-physics software packages in order to support the design and simulation of physical devices and phenomena [38]. The physical equations are discretized on a mesh. The FEM primary advantage is the use of a mesh, which can be variable-sized, with elements of various shapes, making it much better suited to curved geometries. The function of interest $u(r)$ is expanded in terms of basic functions (or "shape functions") tailored to the mesh, $\{\varphi_i(\bar{r})\}$, $u(\bar{r}) = \sum_{i=1}^{N} u_i \varphi_i(\bar{r})$. The wave equation becomes a system of equation for u_i. A solution is achieved using direct or iterative linear and non-linear solvers. The heart of any nonlinear solver, whether in Matlab, Comsol, or elsewhere is some version of the Newton–Raphson iterations: at every stage, the derivative is used to estimate the distance to the solution. The algorithm continues until the error converges below some minimal value.

2.3. Boundary Conditions and Symmetries

The final stage in building a simulation is the choice of boundary conditions. The appropriate choice of boundary conditions implements symmetries which can reduce the domain-size. Mirror symmetries, implemented by reflecting boundary conditions, cut the domain-size in half. Periodic boundary conditions implement a discrete (lattice) translation symmetry, which allows the simulation of an infinite array. They can also be used to implement a discrete rotation symmetry, thereby reducing the domain-size by some finite fraction—a third, a fourth etc. Continuous symmetries, such as axial—i.e., translational—symmetry, and cylindrical—i.e., rotational—symmetry can reduce a three-dimensional simulation to two dimensions. Open or absorbing boundary conditions allow the simulation of infinite domains. "Scattering Boundary Conditions" and "Impedance Boundary Conditions" are well-known varieties; "Port Boundary Conditions" are a proprietary type implemented in Comsol. An alternative method makes use of a region of highly dissipative propagation known as a "Perfectly Matched Layer" (PML). A similar method, known as an Infinite Domain, makes use of a non-linear spatial transformation. Very thin layers—regions of high aspect ratio—are a computational obstacle insofar as they require very fine meshes. This can be avoided, and good accuracy can be obtained by replacing the region with boundary conditions which relate the fields on either side by extrapolation. This is computed based on the material characteristic of the layer—the electrical resistance, thermal conductance, optical transmittance, etc. [32].

3. Analytical Method: Models and Properties of Metallic Nano-Particles

3.1. Analytical Method: Models and Properties of Metallic Nano-Particles

An electromagnetic source excites the nanostructure, and by observing the absorption and scattering cross-section, its electromagnetic properties can be determined. When an impinging electromagnetic wave with an appropriate incident wavelength illuminates a metallic nano-particle, the metals' free electrons start oscillating collectively. Such oscillations lead to the propagation of strong surface waves [39,40], also known as Propagating Surface Plasmon Polaritons (PSPP). The resonant optical properties of nano-particles can be studied, starting with their polarizability expressions. The polarizability value strongly depends on the nano-particle geometry, particularly on its size, shape, inclusion composition, and the surrounding dielectric environment refractive index.

3.2. Electrostatic Approximation and Mie Theory for Metallic Sphere

The solution to the electrostatic problem for a sphere is well known [41]. The electric field solution inside the sphere is given by:

$$E_{In} = \frac{3\varepsilon_e}{\varepsilon_i + 2\varepsilon_e} E_{inc} \tag{1}$$

where E_{In} is the electric field inside the sphere nano-particle, E_{inc} is the incident electric field, ε_e is the surrounding dielectric environment permittivity, ε_i is the inclusion dielectric permittivity. Dimensionless polarizability for the electrostatic problem for a sphere is given by:

$$\beta_s = \frac{\varepsilon_i - \varepsilon_e}{\varepsilon_i + 2\varepsilon_e} \tag{2}$$

The Electro-Static Approximation is very limited because it does not consider size-related effects, therefore the energy conservation is only approximated. Depolarization and radiative corrections to the Electro-Static Approximation (ESA) approach have been made through using Mie theory [1]:

$$\beta_s^{Mie} = \frac{\beta_s}{1 - (K_e a)^2 \left[1 - \frac{2\varepsilon_i + 1}{5(\varepsilon_i - 1)}\right] \beta_s - \frac{2}{3} i (K_e a)^2 \beta_s} \tag{3}$$

where:

- β_s^{Mie} is the corrected polarization by Mie theory;
- β_s is the ESA approximation polarization;
- $K_e = 2\pi n/\lambda$ is the wavenumber in the medium;
- a is the size of the radius of the sphere.

Far field properties such as absorption and scattering cross-section can calculated by using the polarizability β_s^{Mie}:

$$\sigma_{ext} = 4\pi K_e a^3 Im(\beta_s^{Mie}) \tag{4}$$

$$\sigma_{sca} = \frac{8\pi a^2}{3} (K_e a)^4 |\beta_s^{Mie}|^2 \tag{5}$$

The extinction cross-section is the sum of the scattering and absorption cross-section. Therefore, the absorption cross-section is:

$$\sigma_{abs} = \sigma_{ext} - \sigma_{sca} \tag{6}$$

Scattering and absorption cross-section provides great insight for the study of electromagnetic properties of a nano-particle. However, the following assumptions must be made:

- The resonant behavior of the individual nano-particle can be studied in terms of quasi-static approximation; therefore, the size of a nano-particle must be smaller than the wavelength of the light source [42]. The electromagnetic field is approximately constant over the particle volume for small nano-particles.
- The nano-particle macroscopic electromagnetic behavior can be related to its polarizability only if the considered particle is homogeneous, and the surrounding material is a homogeneous, isotropic, and non-absorbing medium.

3.3. Analytical Models of Prolate Spheroid Nano-Particles

The analytical models of prolate spheroid nano-particles, which link the electromagnetic nano-particle properties to their geometrical and structural parameters, are presented in order to describe their resonant behavior. Considering the electric field $= E_0\hat{z} = -\nabla\varphi_0$, the solution for ESA problem of prolate spheroid:

$$\varphi_{in} = \frac{\varphi_0}{1 + L_3\frac{\varepsilon_i - \varepsilon_e}{\varepsilon_e}} = \frac{3\varepsilon_e}{3L_3\varepsilon_i + \varepsilon_e(3 - 3L_3)}\varphi_0 \tag{7}$$

As was discussed for the sphere nano-particle, the dimensionless polarizability for excitation along the Z axis is:

$$\beta_s = \frac{\varepsilon_i - \varepsilon_e}{3L_3\varepsilon_i + \varepsilon_e(3 - 3L_3)} \tag{8}$$

where L_3 is the depolarization factor and is calculated by the following equation:

$$L_3 = \frac{1}{2e^2}\left[1 - \frac{1 - e^2}{2e}\ln\left(\frac{1 + e}{1 - e}\right)\right] \tag{9}$$

$$e = \sqrt{1 - \left(\frac{b}{a}\right)^2} \tag{10}$$

The L_3 factor of a nano-particle plays a crucial role in the polarizability's resonant behavior for the enhancement of localized surface plasmon resonance (LSPR) strength. The far field properties are obtained by using the dipolar approximation. The extinction cross-section equation:

$$\sigma_{ext} = 4\pi K_e abc Im(\beta_s) \tag{11}$$

$$\sigma_{sca} = \frac{8\pi(abc)^2}{3}(K_e)^4|\beta_s|^2 \tag{12}$$

where a, b and c are the spheroid axes, and the absorption cross-section is the same as Equation (6). A Matlab code was used to plot the graphs of the extinction cross-section for silver sphere nano-particle (L $\approx 1/3$), as presented in Figures 1–3.

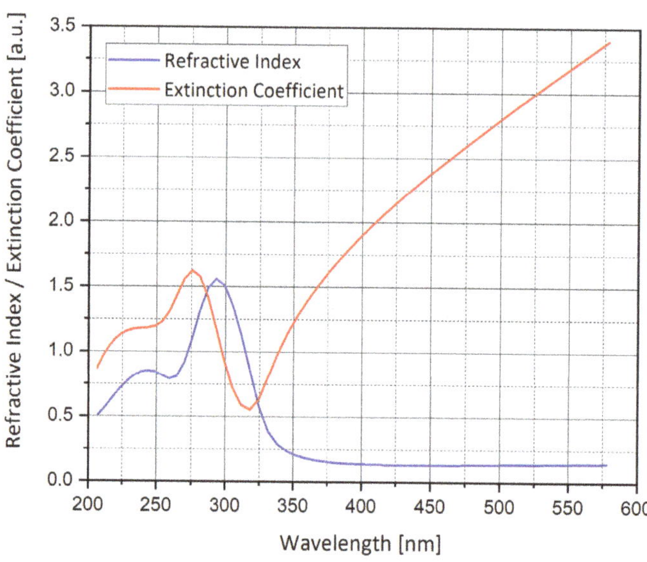

Figure 1. Refractive index and extinction coefficient of silver (Ag).

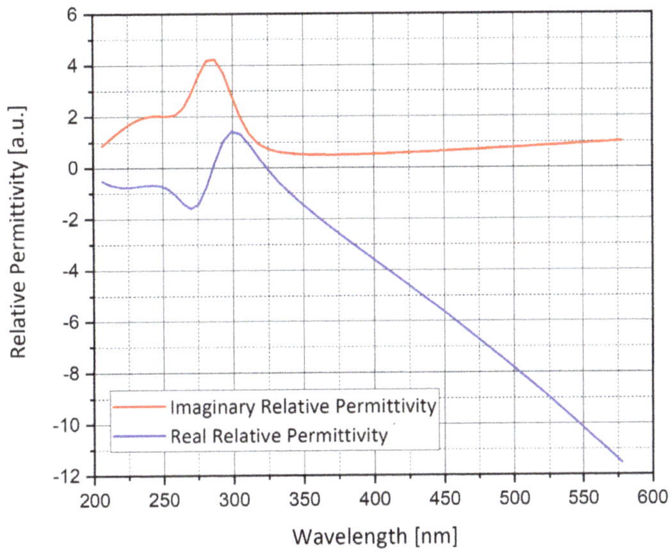

Figure 2. Real and imaginary part of the relative permittivity of silver (Ag).

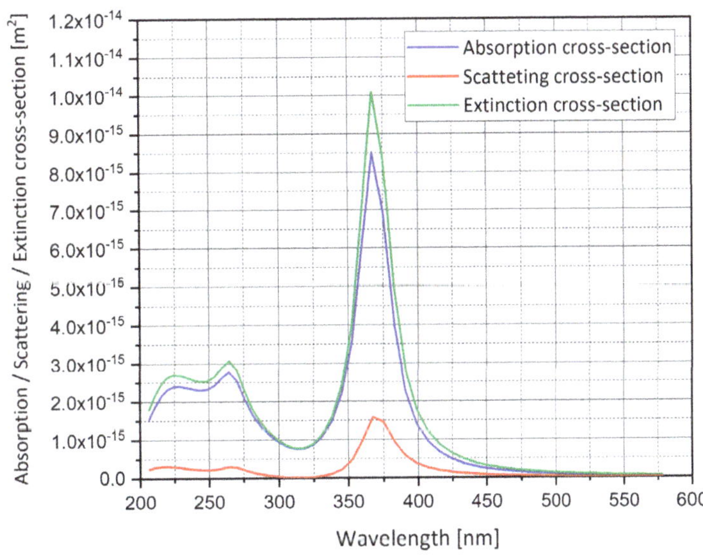

Figure 3. Absorption, scattering and extinction cross-section of silver (Ag) spheroid nano-particle.

Figures 1–3 show the analytical model for silver nano-particle extinction cross-section. Absorption cross-section is higher than scattering cross-section for a small nano-particle, because as a particle grows it becomes closer to the size of the light wavelength, therefore scattering interaction occurs more often compared to a small particle which absorbs the photon and dissipate the photon energy as heat. Absorption and scattering cross-section are competing phenomena and have different application and measurement techniques.

Figure 4 compares the analytical model with the numerical model of a silver sphere nano-particle and shows the difference between sphere geometry and hemi-sphere geometry. The analytical and numerical models matched almost completely. Moreover, the hemi-sphere geometry follows the same pattern as the sphere geometry but the values of the extinction cross-section for the hemi-sphere are smaller.

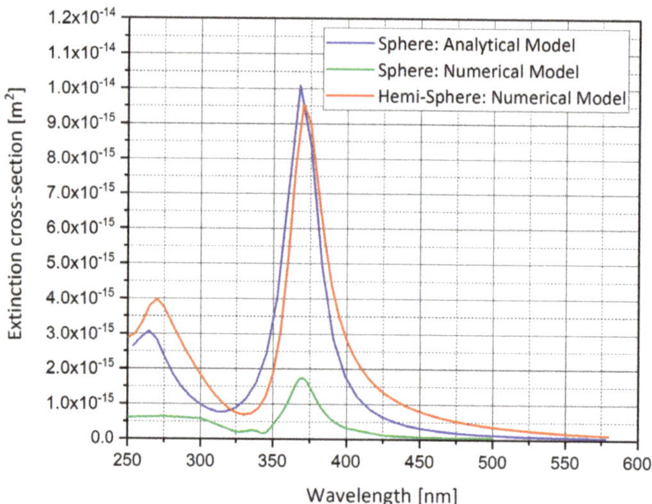

Figure 4. Analytical and numerical model comparison for extinction cross-section of silver (Ag) sphere and hemi-sphere.

Figure 5 compares the analytical model with the numerical model of the silver spheroid nano-particle and shows the difference between spheroid geometry and hemi-spheroid geometry. The analytical and numerical models matched almost completely, moreover the hemi-spheroid geometry follows the pattern as the sphere geometry but the extinction cross-section values for the hemi-spheroid are smaller and the peak is slightly off.

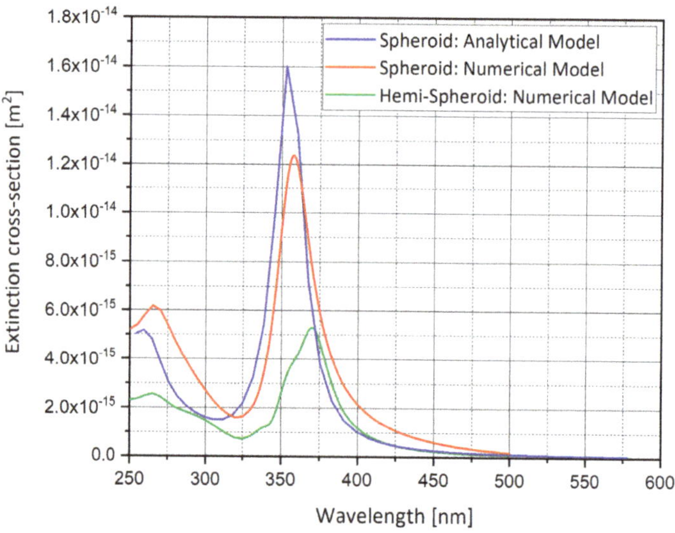

Figure 5. Analytical and numerical model comparison for extinction cross-section of silver (Ag) spheroid and hemi-spheroid.

4. Simulation Results

4.1. Field Enhancement Factor (FFF)

The field enhancement factor for a silver sphere (Figure 6) was calculated by numerical simulation to validate the method of simulation. When comparing the enhancement obtained by a nanostructure and a nano-cavity of the same shape, a curious duality is noted between particles and cavities which exchanges the roles of prolate and oblate spheroids, and between the major and minor axes of any particular spheroid; it is significant in choosing the optimal shape for a given excitation polarization and vice versa. Stratified nanostructures—nano-shells—introduce some freedom in tuning the resonant frequency; the predictions of the Electro-static Approximation (ESA) will be compared to the propagating simulation for the case of spheres. A simplified model of the actual device was studied, by simulating a system of two structures and studying their mutual influence as a function of separation. Following this, a pixel design based on a finite array of nanostructures was studied.

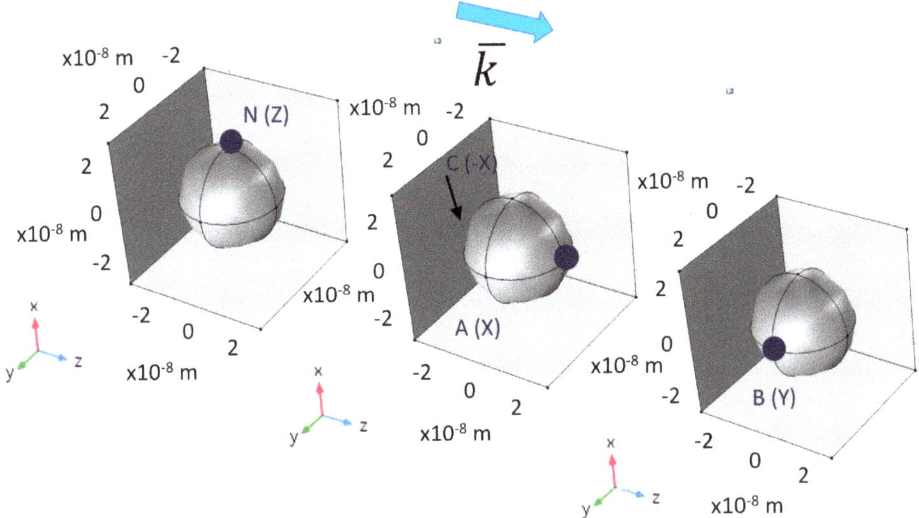

Figure 6. Illustration example of Comsol simulation of the points (N), (A), and (B). The field enhancement is displayed at the points (N), (A), and (B) and some of their polar pairs of the nano-spheres.

4.2. Parameters, Operators and Variables

System constants are necessary for constructing the CAD geometry and values needed in parametric sweeps were defined as parameters (depicted in Table 1). Values that needed to be changed throughout the simulation were defined as variables in a way that made realizing the complex expression more manageable, Table 2.

Table 1. Parameters for all geometries and parametric sweeps.

Name	Expression	Description
W	150 nm, 250 nm, 450 nm	Width of physical geometry
t_pml	30 nm	Perfectly Matched Layer (PML) thickness
h_air	80 nm	Air domain height
h_subs	50 nm	Substrate domain height
T	1.433–8.303 nm	Nano-shell thickness (using Δ = 0.2, 0.4, 0.6, 0.8)
R	20 nm	Nano-particle radius
A	20 nm	Ellipsoid x semi axis
B	20 nm	Ellipsoid y semi axis
C	40 nm	Ellipsoid z semi axis
E	0.866	Eccentricity
Na	1	Air refractive index
Phi	$0, \pi/2$	Azimuthal angle of incidence
θ	$0, \pi/6, \pi/4, \pi/3$	Polar angle of incident field
I_0	10^6 W/m^2	Intensity of incident field
P	$I_0 w^2 \cos(\theta)$	Port power
Sep	5–315 nm	Separation between particles

Table 2. Variables and functions used during the simulations.

Name	Expression	Unit	Description	Domain
ewfd.Ex	0	V/m	X direction electric field	PML Domain
ewfd.Ey	0	V/m	Y direction electric field	PML Domain
ewfd.Ez	0	V/m	Z direction electric field	PML Domain
E0x	-sin(phi)		Amplitude of Ex in X	Port 1,2
E0y	cos(phi)		Amplitude of Ey in Y	Port 1,2
intop_surf			Surface integral	nano-particle surface
intop_vol			Volume integral	nano-particle volume
nrelPoav	n_x * ewfd2.relPoavx + n_y * ewfd2.relPoavy + n_z * ewfd2.relPoavz	W/m^2	Relative normal Poynting flux	Entire model
Sigma_sc	intop_surf(nrelPoav)/I_0	m^2	scattering cross-section	Entire model
Sigma_abs	intop_vol(ewfd2.Qh)/I_0	m^2	absorption cross-section	Entire model
Sigma_ext	Sigma_sc+ Sigma_abs	m^2	extinction cross-section	Entire model

4.3. Geometric Structures, Physics Definitions, Materials and Mesh

The geometric structures checked in the simulations are described in the following figures: hemi-sphere (Figure 7a), hemi-spheroid (Figure 7b), cavity (Figure 7c), nano-cone (Figure 7d), ellipsoidal rod (Figure 7e), ellipsoidal cavity (Figure 7f), double nano-cone (Figure 7g). Multi-tips arrays are also presented in Figure 7h. The Wave Optics module was used in all simulations by using the electromagnetic waves frequency domain (ewfd) model. In the first step of the simulation, the full field was simulated in the physical domain as shown in Figure 7i, and in the second step the scattered field was simulated in all domains. A periodic boundary condition was used, and PML as shown in Figure 7j. The input electric field excitation enters from the top (Figure 7i). The materials used in the simulations, are silver (Ag rakic model) for substrate and nano-particle (Figure 7a), and air for all other domains.

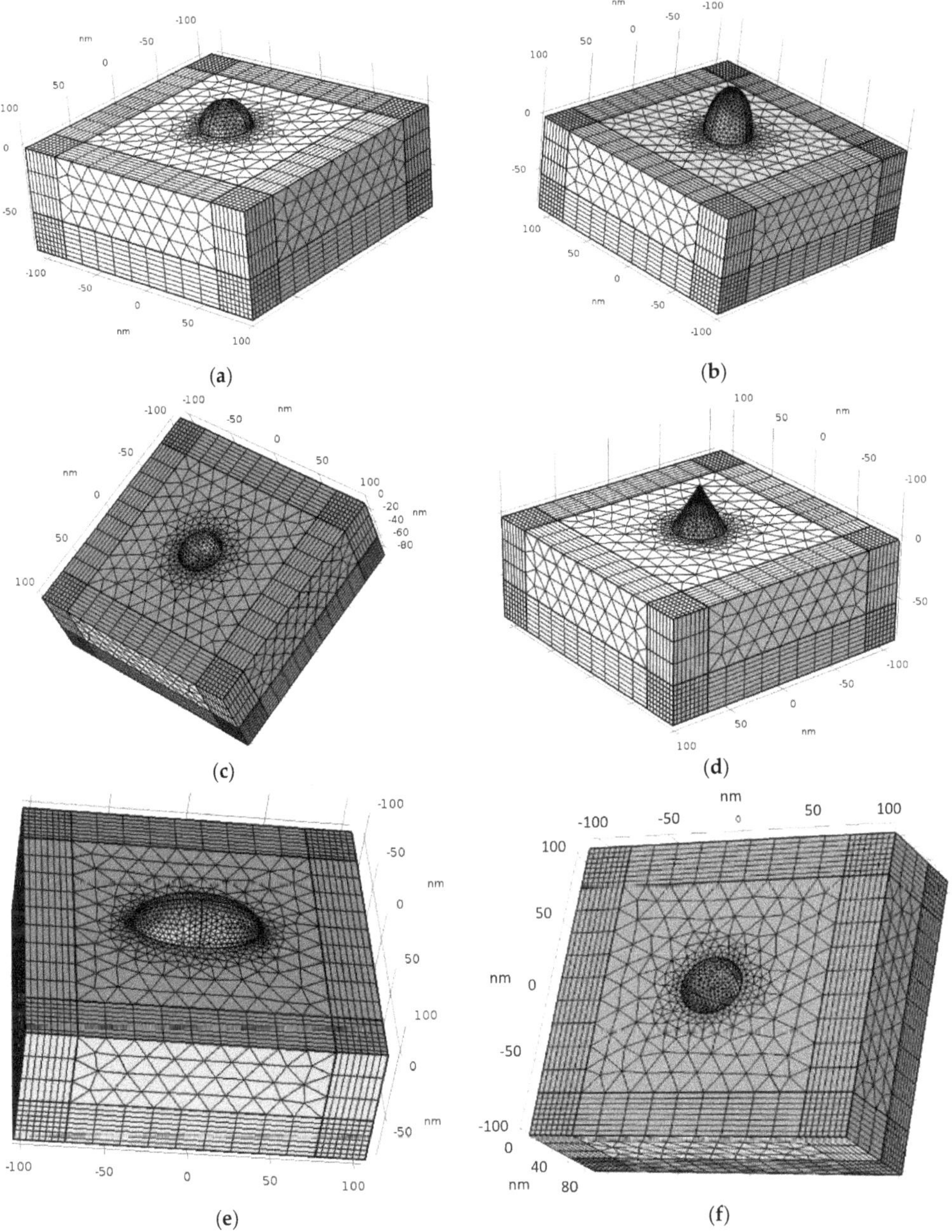

Figure 7. *Cont.*

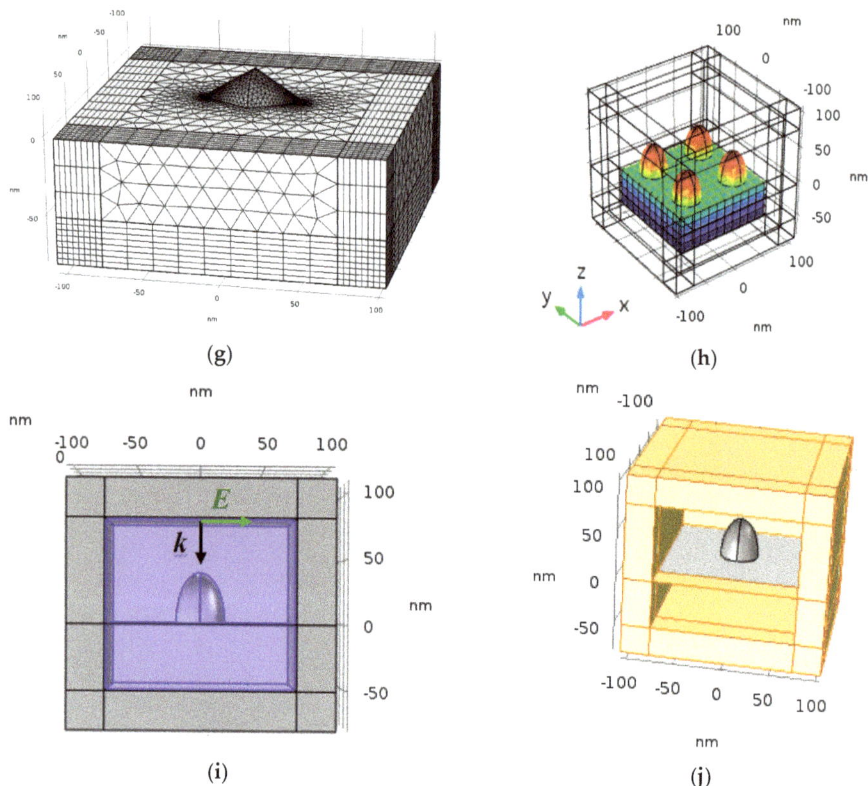

Figure 7. Tip structure different geometries: (**a**) hemi-sphere nano-particle; (**b**) hemi-spheroid nano-particle; (**c**) cavity; (**d**) nano-cone; (**e**) ellipsoidal rod; (**f**) ellipsoidal cavity; (**g**) double nano-cone; (**h**) SERS/TERS square; (**i**) the physical domain and the port input electric field excitation; (**j**) the PML domain.

4.4. Solvers and Studies

In order to identify the optimal shape and geometry, seven shape studies were performed: hemi-sphere, cavity, hemi-spheroid, nano-cone, ellipsoidal cavity, ellipsoidal rod and double nano-cone. The simulations consist of two steps: simulation of the full field (ewfd) and simulation of the scattered field (ewfd2). All the simulations used an input electrical field with a wavelength in the range of 250–500 nm. While the width of physical geometry W = 150 nm and a parametric sweep for $\theta = 0, \pi/6, \pi/4$, and $\pi/3$ were, respectively, used in the simulations of the hemi-sphere, cavity, hemi-spheroid, and ellipsoidal cavity (with the exception of $\theta = \pi/4$). Following are the results.

4.5. Hemi-Sphere Geometry Results

By impinging light in the Z axis, with electric field polarized in the Y axis, with different polar angles $\theta = 0, \pi/6, \pi/4$, and $\pi/3$ and in different wavelengths, the E^4 approximation and the extinction cross-section can be calculated as, respectively, shown in Figure 8a,b. As the polar angle θ gets bigger, the enhancement factor gets smaller. The peak enhancement is at 368 nm. When the K vector of the input electric field is normal to the substrate with the nano-particle the field enhancement is largest. The extinction cross-section shows the same behavior when changing the polar angle θ as it was in the E^4 calculation above. The peak is at 368 nm, and at $\theta = 0$ is the largest extinction cross-section.

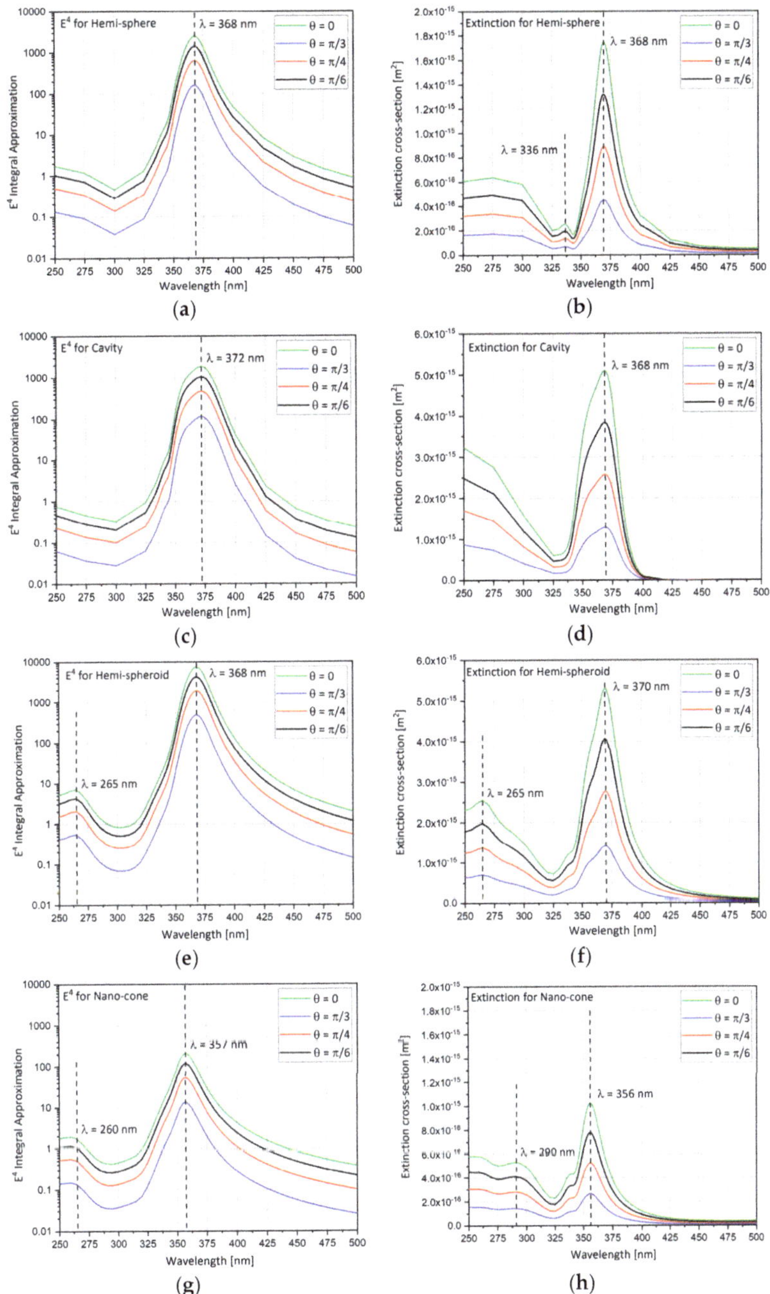

Figure 8. E^4 approximation and extinction cross-section for changing polar angle θ in wavelengths range of λ = 250–500 nm. (**a**) E^4 approximation for hemi-sphere; (**b**) extinction for hemi-sphere; (**c**) E^4 approximation for cavity; (**d**) extinction for cavity; (**e**) E^4 approximation for hemi-spheroid; (**f**) extinction for hemi-spheroid; (**g**) E^4 approximation for nano-cone; (**h**) extinction for nano-cone.

4.6. Cavity Geometry Results

As discussed above in the hemi-sphere shape analysis, the E^4 approximation and the extinction cross-section can be calculated (Figure 8c,d). Again, as the polar angle θ gets bigger, the enhancement factor gets smaller. The peak enhancement is at 372 nm. When the K vector of the input electric field is normal to the substrate with the cavity the field enhancement is largest. The extinction cross-section shows the same behavior when changing the polar angle θ as it was in the E^4 calculation above. The peak is at 371 nm, and at θ = 0 the extinction cross-section is the largest.

4.7. Hemi-Spheroid Geometry Results

E^4 approximation and extinction cross-section are calculated (Figure 8e,f). The hemi-spheroid that was used here is with eccentricity of 0.866. As polar angle θ gets bigger, the enhancement factor gets smaller. The peak enhancement is at 368 nm. When the K vector of the input electric field is normal to the substrate with the nano-particle, the field enhancement is largest. The extinction cross-section shows the same behavior when changing the polar angle θ as it was in the E^4 calculation above. The peak is at 368 nm, and at θ = 0 the extinction cross-section is the largest.

4.8. Nano-Cone Geometry Results

The E^4 approximation and the Extinction cross-section are presented (Figure 8g,h). The nano-cone surface area is the same as the hemi-sphere surface area. As polar angle θ gets bigger, the enhancement factor gets smaller. The peak enhancement is at 357 nm. When the K vector of the input electric field is normal to the substrate with the nano-cone, the field enhancement is largest. The extinction cross-section shows the same behavior when changing the polar angle θ as it was in the E^4 calculation above. The peak is at 355nm, and at θ = 0 the extinction cross-section is the largest.

4.9. Ellipsoidal Cavity Geometry Results

In addition to above standard shapes, additional complex configurations were also analyzed, like the ellipsoidal cavity geometry, analyzed here. As discussed in the hemi-sphere, the E^4 approximation and the extinction cross-section can be calculated (Figure 9a,b). As polar angle θ gets bigger, the enhancement factor gets smaller. The peak enhancement is at 375 nm. When the K vector of the input electric field is normal to the substrate with the cavity, the field enhancement is largest. The extinction cross-section shows the same behavior when changing the polar angle θ as it was in the E^4 calculation above. The peak is at 372 nm, and at θ = 0 the extinction cross-section is the largest.

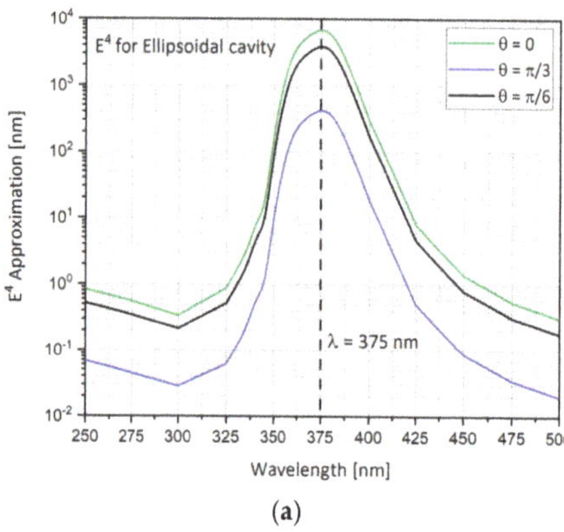

(a)

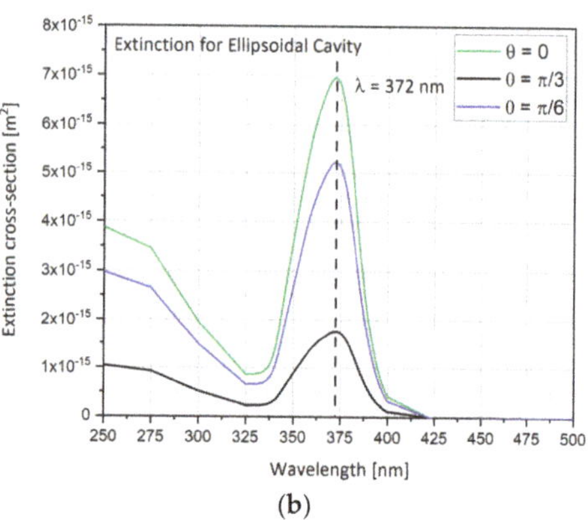

(b)

Figure 9. Ellipsoidal cavity results while changing the polar angle θ in a wavelengths range of λ = 250–500 nm. (**a**) E^4 approximation; (**b**) extinction cross-section.

4.10. Ellipsoidal Rod Geometry Results

The ellipsoidal rod geometry is analyzed in this section. As discussed in the hemisphere the E^4 approximation and the extinction cross-section can be calculated (Figure 10). The ellipsoidal rod that was used here is with eccentricity of 0.866 in the Y direction. Contrary to other geometries, this geometry is sensitive to the electric field polarization. When the electric field polarization is in the Y direction ($\varphi = 0$), localized surface plasmon (LSP) is produced like a dipole in accordance with the electric field polarization. This time, the peak enhancement is at 378 nm. The same happens in the perpendicular direction X for polarized electric field in the X direction ($\varphi = \pi/2$). The peak enhancement is at 363 nm. When the electric field polarization is $\varphi = 0$ the peak that is produced is higher than the peak produced by $\varphi = \pi/2$.

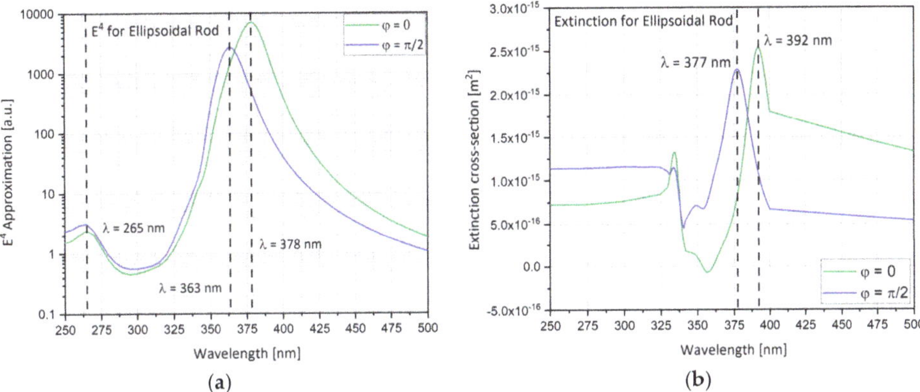

Figure 10. Ellipsoidal rod results while changing the polar angle θ in a wavelengths range of λ = 250–500 nm. (**a**) E^4 approximation; (**b**) extinction cross-section.

4.11. Double Nano-Cone Geometry Results

The double nano-cone geometry is analyzed in this section. As discussed in the hemisphere, the E^4 approximation and the extinction cross-section can be calculated (Figure 11). As discussed in the ellipsoidal rod, this geometry is sensitive to the electric field polarization. When the electric field polarization is in the Y direction ($\varphi = 0$), localized surface plasmon (LSP) is produced like a dipole in accordance with the electric field polarization. The peak enhancement for polarization in Y direction ($\varphi = 0$) is at 370 nm. The same happens in the perpendicular direction X for the polarized electric field in the X direction ($\varphi = \pi/2$). The peak enhancement for polarization in Y direction ($\varphi = \pi/2$) is at 375 nm. When the electric field polarization is $\varphi = \pi/2$, the produced peak is higher than the peak produced by $\varphi = 0$.

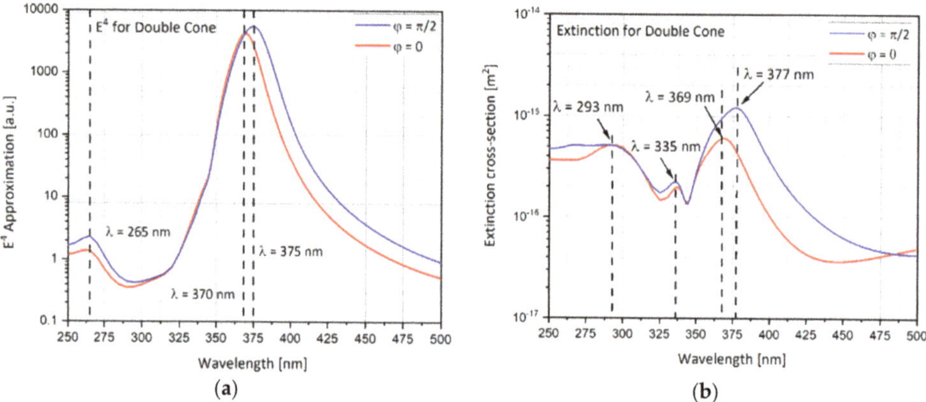

Figure 11. Double nano-cone results while changing the polar angle θ in a wavelengths range of λ = 250–500 nm. (**a**) E^4 approximation; (**b**) extinction cross-section.

4.12. Results Comparison between Different Nano-Particles Geometries

The hemi-sphere, cavity, hemi-spheroid and the nano-cone shapes are now compared. As shown in Figure 12, the spheroid has the largest SERS Enhancement Factor (EF) reaching up to 7300, followed by the sphere with SERS EF of 2500, the cavity with SERS EF around 1800 and finally the nano-cone with SERS EF less than 300. Figure 13 shows the comparison of the extinction cross-section between geometries and the spheroidal cavity and the hemi-spheroid shows the highest extinction cross-section values, but the extinction cross-section is highest in the ellipsoidal cavity by far reaching up to $7 \cdot 10^{-15}$ m^2, as shown from Figure 9b. In order to summarize the seven presented options and to classify them by preference criteria, Table 3 includes the main parameters and obtained values.

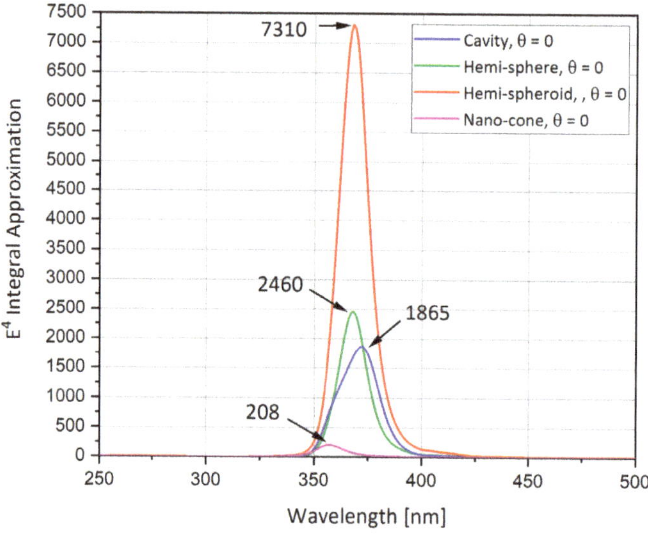

Figure 12. Surface-Enhanced Raman Scattering (SERS) EF: comparison between different nano-particle geometries.

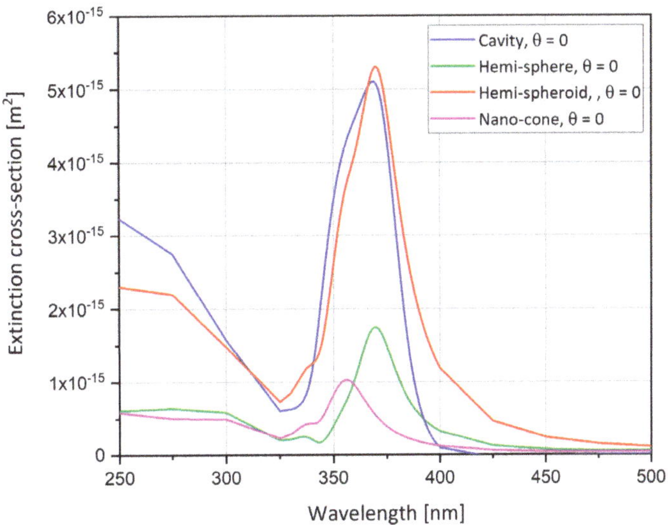

Figure 13. Extinction cross-section: comparison between different nano-particle geometries.

Table 3. Comparison table of the studied particle's shapes based on E^4 approximation.

Studied Shape	Checked Polar or Azimuth Angles	λ Peak Enhancement	E^4 at Peak	Comments
Hemi-sphere	$\theta = 0, \pi/6, \pi/4, \pi/3$	λ = 368 nm	2460	• Largest extinction cross-section at θ = 0
Spheroidal cavity	$\theta = 0, \pi/6, \pi/4, \pi/3$	λ = 372 nm	1865	• Largest extinction cross-section at θ = 0. • Easiest shape using Focused Ion Beam (FIB).
Hemi-spheroid	$\theta = 0, \pi/6, \pi/4, \pi/3$	λ = 368 nm	7310	• Largest extinction cross-section at θ = 0. • Eccentricity of 0.866. • Highest field enhancement.
Nano-cone	$\theta = 0, \pi/6, \pi/4, \pi/3$	λ = 357nm	208	• Largest extinction cross-section at θ = 0. • Nano-cone surface area = hemi-sphere's area.
Ellipsoidal cavity	$\theta = 0, \pi/6, \pi/3$	λ = 375nm	6832	• Largest extinction cross-section at θ = 0. • Easiest feasible shape using FIB. • Highest extinction cross-section.
Ellipsoidal rod	$\theta = 0,\ \varphi = 0,\ \pi/2$	λ = 378nm	7038	• Largest peak at $\varphi = 0$. • Sensitive to electric field polarization
Double nano-cone	$\theta = 0,\ \varphi = 0,\ \pi/2$	λ = 375nm	5446	• Largest peak at $\varphi = \frac{\pi}{2}$. • Sensitive to electric field polarization

4.13. Silver vs. Gold vs. Aluminum

As part of the optimization process, the identification of the tip material was also investigated. The simulations compare between silver (Ag), gold (Au) and aluminum (Al). The goal was to determine which material is more suited for higher SERS EF. As shown in Figures 14 and 15, silver/gold/aluminum nano-sphere with radius of 20 nm is simulated. The silver nano-sphere produces the highest peak of electric field enhancement, then the aluminum nano-sphere and lastly the gold nano-sphere, but the peaks are shown to be at different wavelengths. Silver seems to be more suitable for field enhancement in the region of wavelengths of 325–495 nm as shown in Figure 7.

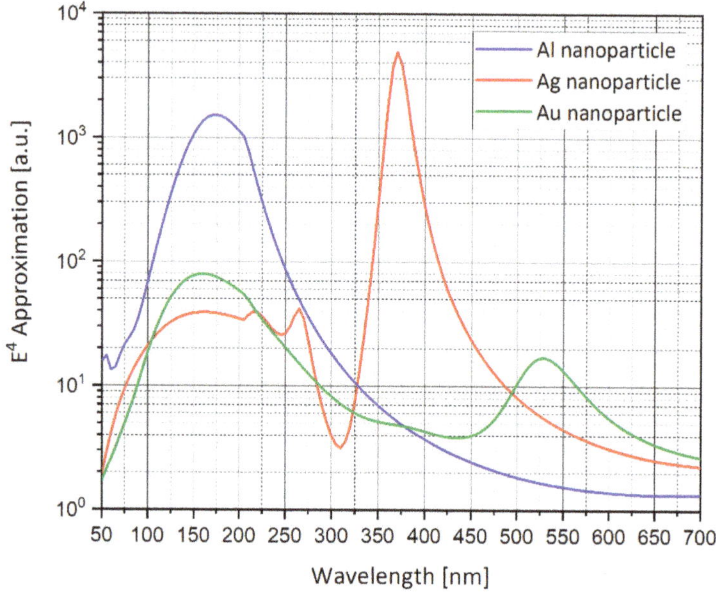

Figure 14. Field Enhancement vs. wavelength for an incident plane wave on a silver (Ag), gold (Au) and aluminum (Al) 20 nm-radius sphere. The E^4 approximation for the electric field enhancement is displayed. Relevant wavelength peaks: silver at λ = 370 nm, gold at λ = 530 nm, aluminum at λ = 175 nm.

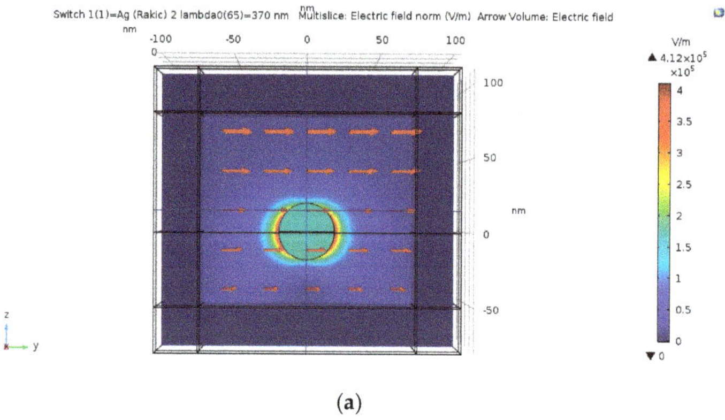

(**a**)

Figure 15. *Cont.*

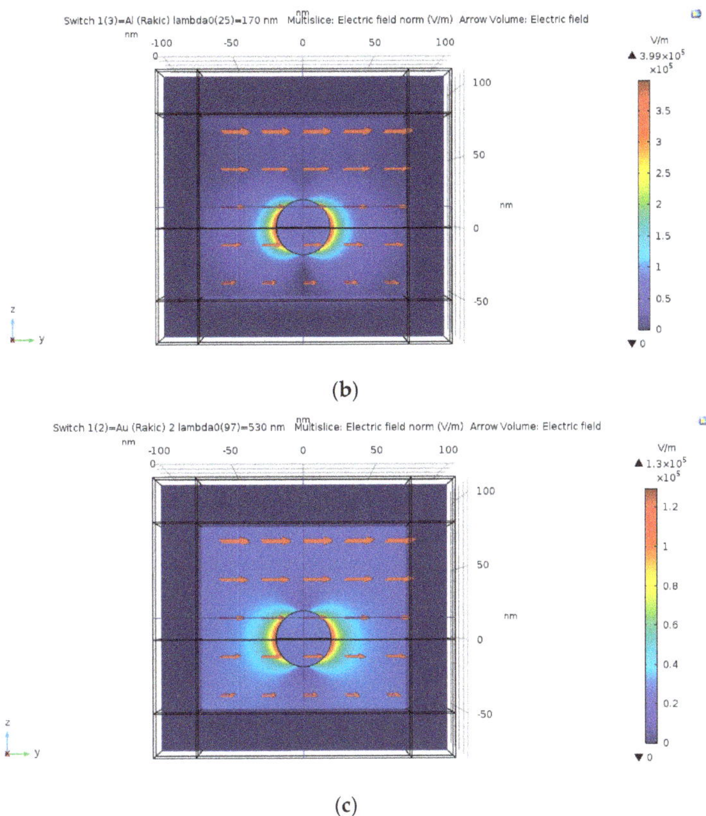

(**b**)

(**c**)

Figure 15. Simulations of silver (**a**), aluminum (**b**) and gold (**c**) nano-particle with the input of electric field in the Y direction and with k vector in the -Z direction.

4.14. Nano-Shells Tuning

Stratified nanostructures—nano-shells, first introduced by (Halas, 2005) [43]—are seen to introduce some freedom in tuning the resonant frequency (Figures 16 and 17). A silver nano-shell with an outer radius (R) of 20 nm and inner radius (r) 11.697–18.567 nm is simulated.

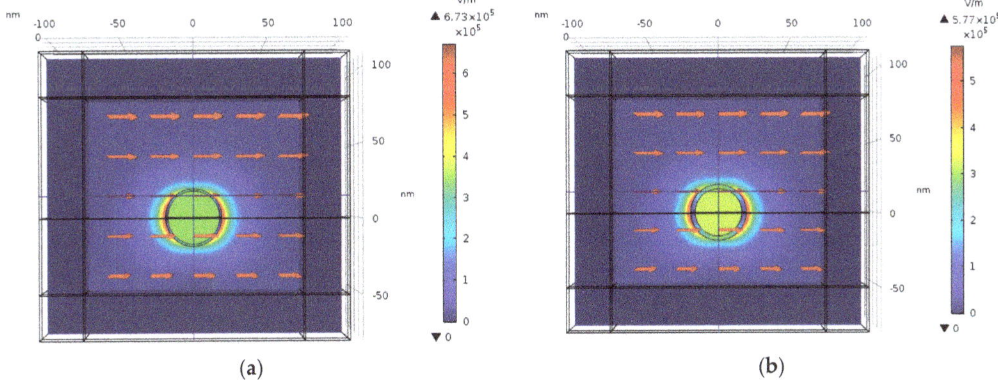

(**a**) (**b**)

Figure 16. *Cont.*

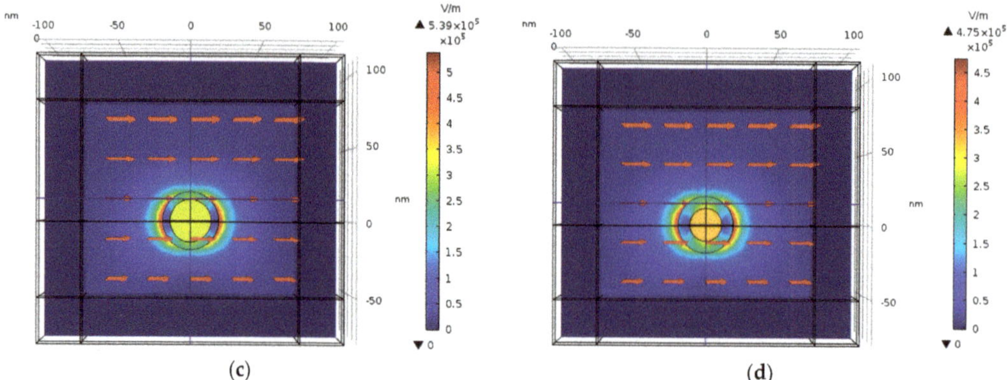

Figure 16. Simulation of nano-shells with thickness (T) ranging from 1.433 to 8.303 nm. The impinging electric field is polarized in the Y direction as shown by the red arrows and the wave front direction (K vector) is in the -Z direction. Δ = 0.2 (**a**), Δ = 0.4 (**b**), Δ = 0.6 (**c**), and Δ = 0.8 (**d**), where $\Delta = 1-(r/R)^3$.

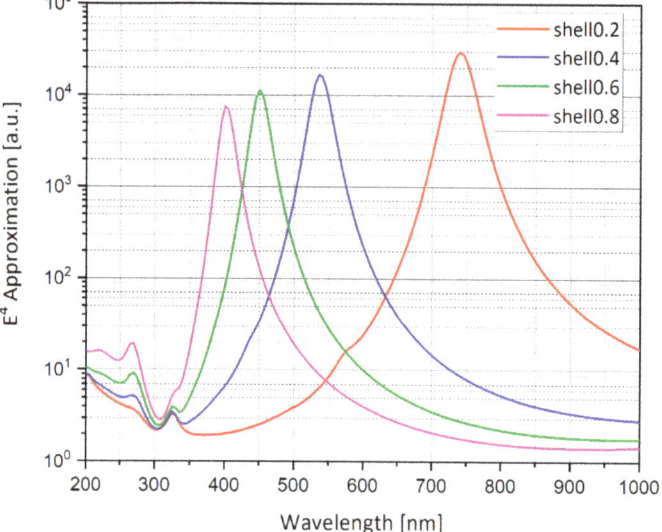

Figure 17. Enhancement curves for spherical nano-shells with outer radius (R) of 20 nm and inner radius (r) such that the nano-shell thickness is T = R − r. The nano-shells thickness ranges between T(Δ = 0.2) = 1.433 nm and T(Δ = 0.8) = 8.303 nm. The shells are excited by a plane wave. The resonance peak shifts depending on the shell thickness. Moreover, the electric field enhancement gets bigger as the nano-shell thickness decreases.

As shown in Figure 17, the nano-shell thickness T = R − r provides a freedom in tuning the resonant wavelength. Moreover, as the shell thickness becomes smaller, the electric field enhancement grows. The sphere has external radius R, internal radius r, and hence thickness R − r. In the ESA, the field enhancement at the North Pole (N) is:

$$M = \left| \frac{\epsilon\epsilon_M + \frac{2}{3}(\epsilon - \epsilon_M)\epsilon\Delta}{\epsilon\epsilon_M + 2\left(\frac{\epsilon - \epsilon_M}{3}\right)^2\Delta} \right|^2 \tag{13}$$

where:

$$\Delta = \left(1 - \frac{r^3}{R^3}\right) \tag{14}$$

The solid sphere corresponds to Δ = 1 while the shell of vanishing thickness is described by Δ = 0. In the latter case $M \rightarrow 1$, as consistency demands. Resonance occurs at when (13) is maximal. For a solid structure, the resonance is achieved for a particular wavelength, determined by the form of $\epsilon(\lambda)$. By contrast, expression (13) can be maximized for any value of λ by setting appropriate Δ. Thus, one may choose a convenient wavelength and achieve resonance by tuning the thickness of the shell. This analytical model is compared to the E^4 approximation calculation in the numerical simulation of nano-shells as shown in Figure 18. In the simulation, Δ = 0.2, 0.4, 0.6 and 0.8, respectively, as shown in Figures 16 and 17.

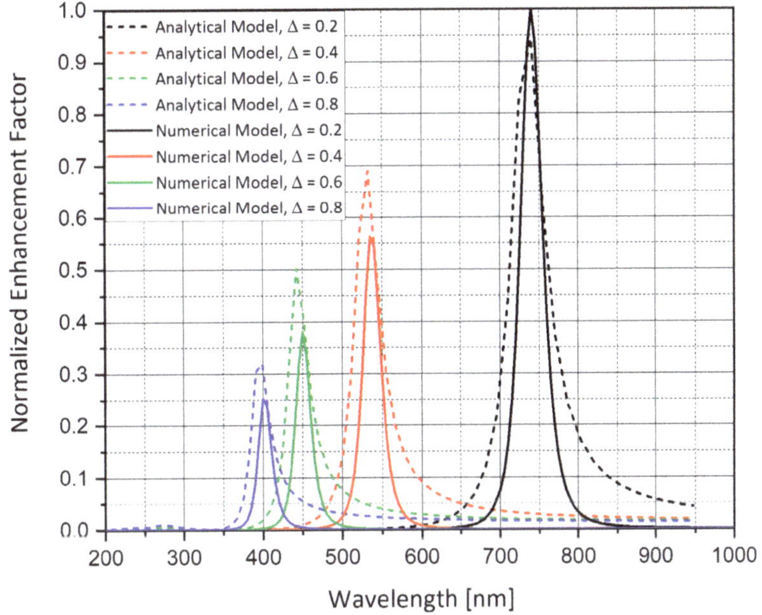

Figure 18. Analytical and numerical models for the normalized enhancement factor.

As shown in Figure 18, the analytical and numerical calculations for the normalized enhancement factor are matched on the resonance wavelength.

4.15. Multiple Nanostructures Mutual Influence

4.15.1. The Influence of Separation between Nanostructures

A simplified model of the actual device was studied by simulating a system of four structures in a box with the width of physical geometry W = 450 nm (Figure 19) and studying their mutual influence as a function of separation (Figure 20). This is the E^4 approximation vs. the separation.

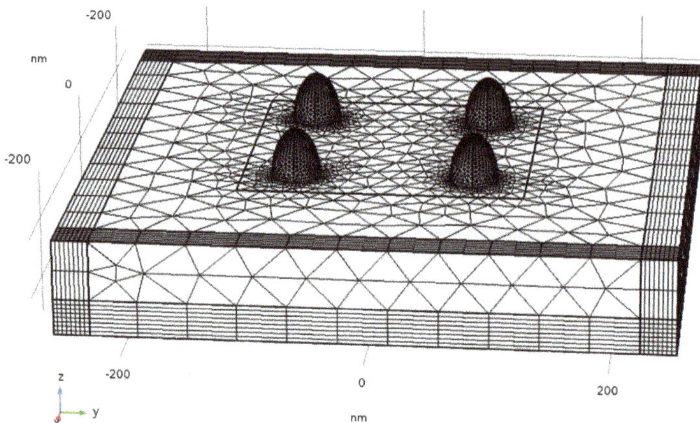

Figure 19. Simulation configuration of neighboring silver hemi-spheroid.

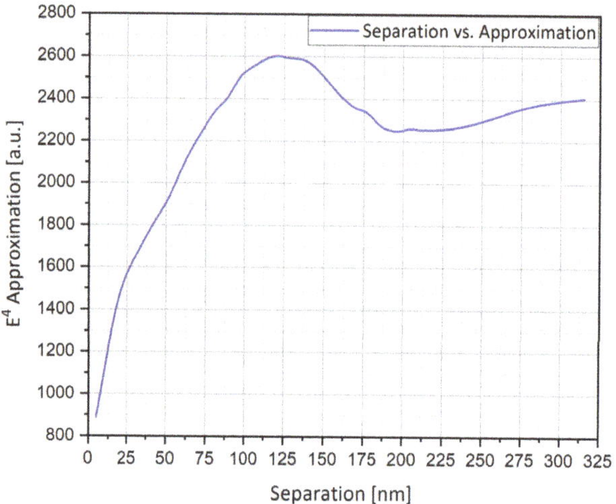

Figure 20. Enhancement vs. separation, for four silver hemi-spheroid in a plane wave, polarized orthogonally to axis of separation. Suppression is observed at short distances. At ~110 nm, the radiation field dominates the localized surface plasmon (LSP) (near) field.

The mutual influence of neighboring nanostructures was investigated numerically, by following the total integrated enhancement as a function of the separation. Figure 21 displays the extinction cross-section from four particles of silver (Ag) hemi-spheroid subject to oscillating electric field; one clearly discerns that the graph is leveled for large distances and starts decreasing at smaller distances, beneath ~100 nm (there is no maximum because the number of structures is constant). The mutual influence is thus negligible at micrometric distances—the order of a pixel—while at nanometric separation it becomes significant.

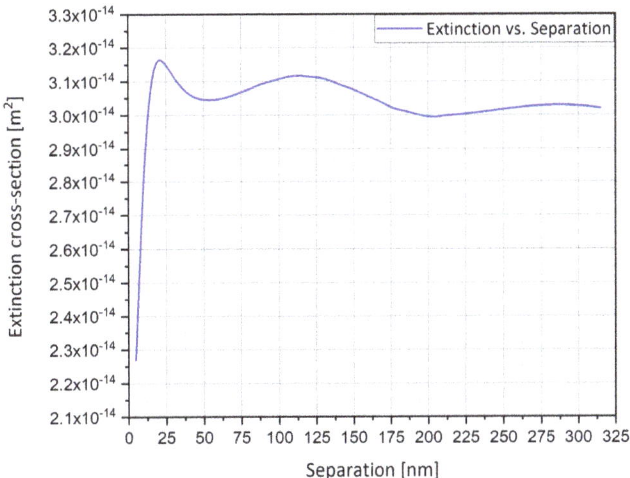

Figure 21. Extinction cross-section vs. separation for four silver spheres in an oscillating electric field in the Y direction.

The appearance of a maximum (in Figure 20) at ~110 nm may seem surprising. However, the expectations outlined above were based on the ESA. In practice, the field of a dynamic radiating sphere is like that of a dipole (Figure 22). It includes several terms, particularly the local field—this is the LSP field and is the only term seen in the ESA, i.e., by a uniform field. This term causes suppression. It varies ~$1/r^3$, the inverse cube of the separation. It also includes the radiation field—this term is only excited by an oscillating field. On the equator, the radiation field is parallel to the source dipole; hence, it causes mutual enhancement, as illustrated in Figures 20 and 22. This term decays as $1/r$. Competing phenomena of mutual dipole suppression and enhancement and—at distances smaller than ~5 nm—of gap or hybrid plasmons [1], lead to an optimal separation for maximal total enhancement.

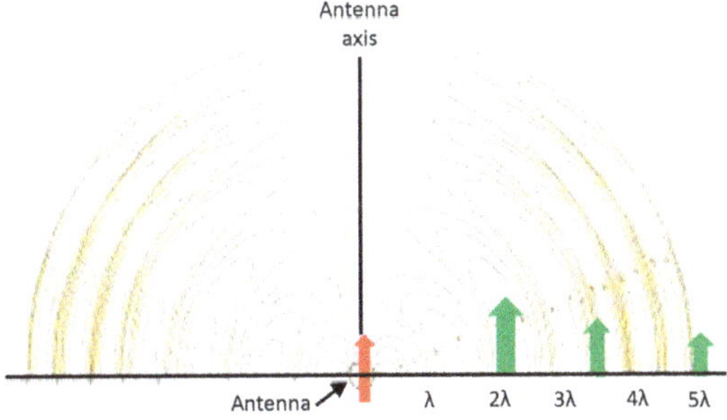

Figure 22. The radiation pattern of a dipole oscillator. The field on the axis is parallel to the dipole leading to enhancement.

4.15.2. The Comparison between Preliminary Results for the Optimal Separation

Preliminary results, as shown in Figures 20 and 21, present two different optimal separations of nano-particles:

- At a separation of ~110 nm (Figure 20), the highest field enhancement is produced.
- At a separation of ~20 nm (Figure 21), the highest extinction cross-section is produced.

These separations are compared in Figures 23 and 24. In these simulations, the width of the physical geometry W = 250 nm.

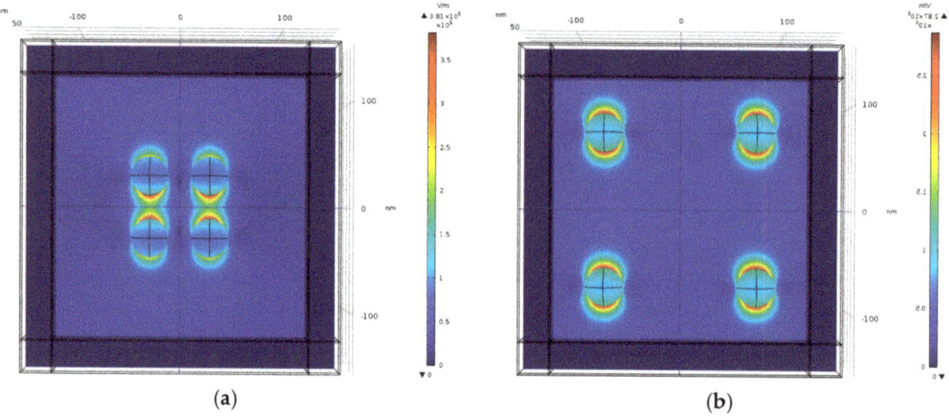

Figure 23. Simulation of separated hemi-spheroid nano-particles. (a) Separation of 20 nm; (b) separation of 110 nm. The width of physical geometry W = 250 nm. The electric field polarization is in the Y direction, therefore the localized surface plasmon is excited mostly in that direction. There is less interaction between nano-particles separated in the X direction.

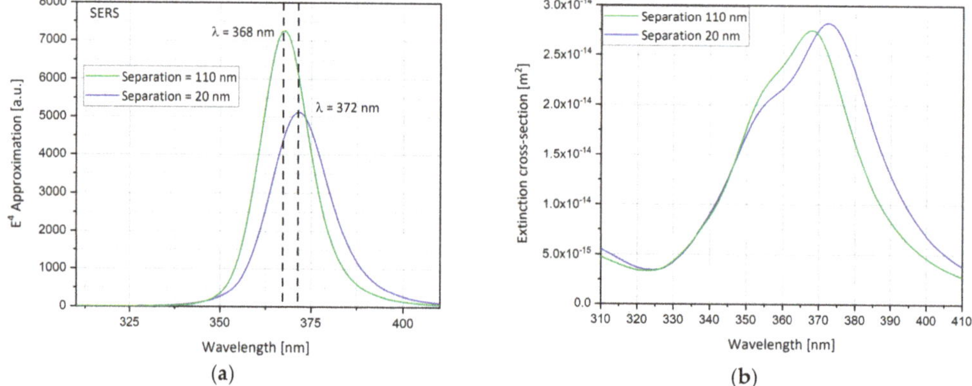

Figure 24. Simulation of four separated hemi-spheroid particles (separation = 20 nm, 110 nm). The width of physical geometry W = 250 nm. (a) E^4 approximation; (b) extinction cross-section.

As depicted in Figure 24a, separation of 110 nm produces a higher peak than separation of 20 nm. The nano-particles (Figure 23a) are very close, which cause plasmons to interact with each other, therefore the mutual dipole suppression effect decreases the electric field enhancement. However, as the separation gets bigger this plasmon interaction gets weaker (Figure 23b), therefore mutual dipole enhancement produces higher electric field enhancement and reaches an optimal separation around 110 nm. The peaks of the field enhancement are produced at different wavelengths. The wavelength for the resonant condition is red-shifted as the gap between nano-particles gets smaller. Figure 24b shows that the extinction cross-section peak for the separation of 110 nm is lower than the extinction cross-section peak for the separation of 20 nm. However, the peaks are produced at different wavelengths and the difference between them is small, therefore the optimal separation is 110 nm, which also produces higher electric field enhancement as shown in Figure 23a.

A pixel design was devised, based on a finite array of hemi-ellipsoidal silver nanostructures of radius 20 nm and an aspect ratio, A.R. = 2.00, on a silicon substrate. An initial study was conducted to determine the optimal number of structures per pixel, or equivalently—for fixed pixel dimension—the optimal distance between them. The initial design chosen comprises 121 structures arranged in a finite square array of dimension 1.1 μm with 11 structures in each direction—a separation of 110 nm.

Further validation studies are necessary, comparing the results of the simulation to analytical results [1] for a few simple geometries such as the sphere and ellipsoid. A comparison of the performance of cavities, and particles of the same shape, will be examined. The particles and protrusions are expected to show better enhancement than the corresponding cavities. The cavity–particle duality will be verified next. The prediction of the ESA for spherical shells will be compared to the simulation for both static and propagating fields; the shape itself will then be optimized. The hemi-spheroids used in the design possess a sharp edge along the bottom face. The significance of the resulting singular field in particular, and the deviation of the performance from ideal spheres and ellipsoids in general, must be examined.

5. Preliminary Experimental Results

Protrusions and Cavities Arrays Fabrication and Structural Characterization

Following the above numerical and analytical analyses, specifying the definition of the optimal material and geometry of the individual tip-probe of the pixels array, preliminary experimental results were performed. Several arrays of protrusions and cavities were fabricated. The arrays were manufactured using a Focused Ion Beam (FIB) milling equipment, at Bar-Ilan university Institute for Nanotechnology and Advanced materials (BINA). Integrated SEM served to characterize the fabrication and to monitor the quality of the samples. It should be noted that due to the COVID-19 world pandemic, we dealt with major limitations and restrictions on regular laboratory work. A series of additional experiments are scheduled to be held in the near future.

The architecture and design steps required extensive work of optimization until reaching the final array, since there is no significance to a single stand-alone tip-probe for these scanning applications. The following figure presents the design of arrays and masks of nano-cones (Figure 25a) and nano-holes (Figure 25b), the fabrication of arrays of cavities (Figure 25c,d), and of protrusions (Figure 25e,f).

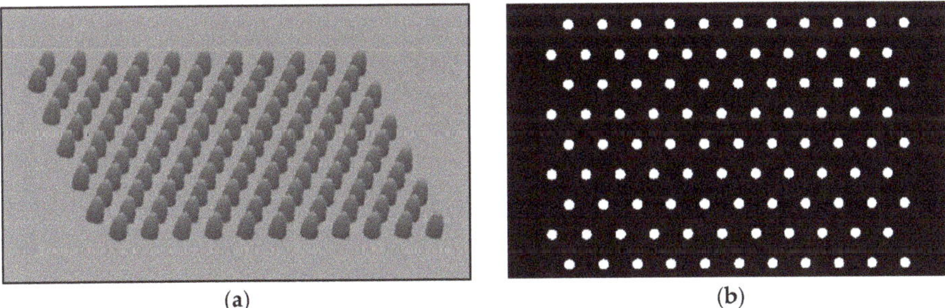

(a) (b)

Figure 25. *Cont.*

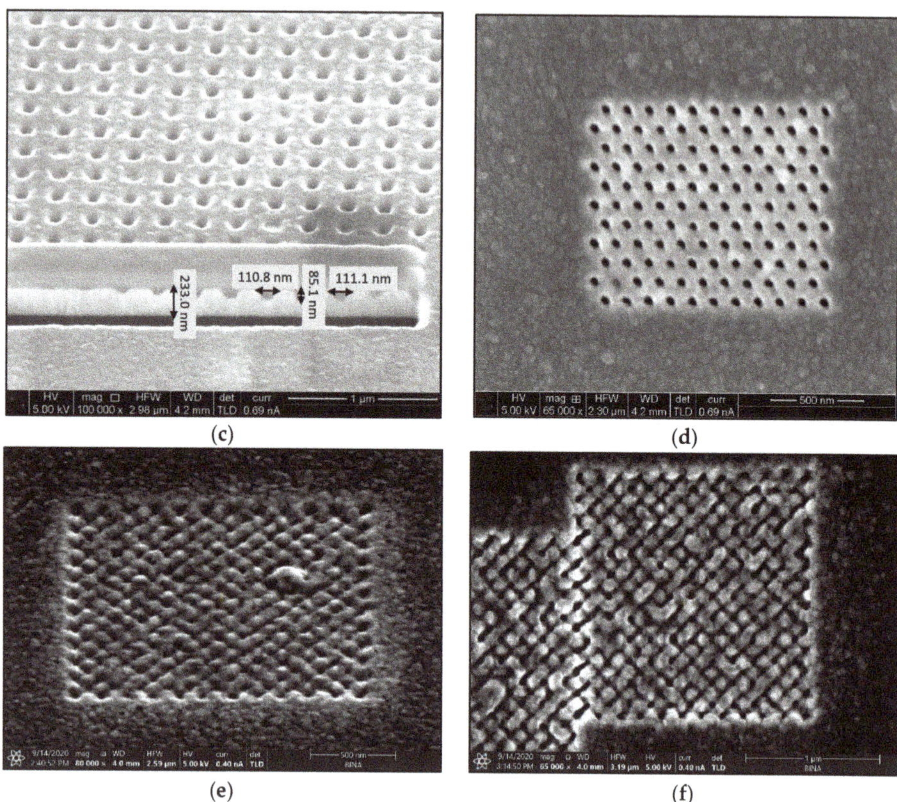

Figure 25. SEM pictures of preliminary samples fabricated with FIB technique. (**a**) Simulation mask of an array of tips before the fabrication of the protrusions; (**b**) simulation mask of holes array before the fabrication of cavities. (**c**) Nano-holes square lattice of cavities; (**d**) nano-holes hexagonal lattice of cavities; (**e**) nano-protrusions array using FIB; (**f**) nano-protrusion double-array fabrication using FIB.

While the protrusions are difficult to obtain, the cavities were obtained in a much more straightforward resolution. Looking at the dimensions of the pixel, and in particular at the total array active area (white space), one can obtain a matrix of width × height = 1300 nm × 1080 nm. In fact, the active area consists of an arrangement of 11 × 11 nanostructures. In a preliminary configuration, the structures are depressions (i.e., open cavities), to be drilled into the silver layer. The opening is circular of radius r = 20 nm. The separation distance between structure centers is 120 nm, so the separation between the structure edges remains 80 nm. Regarding the pixel depth and repetition, the following dimensions were chosen: 10 pixels separated by at least 5 to 10 µm for a good separation in an optical microscope. The structure of the first pixel should have a depth of 20 nm, i.e., it should be semi-spherical. The other pixels should be of increasing depth until a maximum depth of 120 nm. Recording the current and time used for each pattern was crucial in order to determine the plasmonic properties as a function of the ion dose. Pictures of the arrays design and of the preliminary results are presented in Figure 25.

6. Discussion

6.1. The Nano-Particles Geometry

The simulations from previous sections provide an insight to what is the optimal nano-particle geometry. Analyzing Figure 12 and Table 3, one can observe that the hemi-spheroid geometry is the most optimal one for obtaining the highest SERS EF. Looking at Figure 9b, it seems clear that the ellipsoidal cavity provides the highest extinction cross-section. Moreover, the ellipsoidal rod and the double nano-cone simulation provide excellent insight into the effect off polarization of the electric field on specific geometries. The nano-cone simulation produced very low SERS EF, because the polarization of the electric field was not aligned with the tip of the cone. This outcome was verified in the double nano-cone simulation which produced higher SERS EF, because the polarization was aligned with the tips of the double nano-cone geometry.

More research is required in order to ascertain what the optimal eccentricity of the hemi-spheroid should be. Moreover, combining different geometries such as hemi-spheroid with a nano-cone could provide a better SERS EF, or even act as another tool to tune the resonant frequency, as was demonstrated in the nano-shells geometry. For the first generation of a TERS device, the hemi-spheroid geometry is most certainly going to be a good starting point for device measurements, characterization and advancing research in this direction.

6.2. Particles Material and Nano-Shells Tuning

The simulations from Section 4.13 provide further insight into which material should be used for the nano-particles. Silver shows great promise in the wavelength region of 325–495 nm (Figure 14). Silver produces the highest SERS EF and extinction cross-section, but aluminum could be used in the region 50–325 nm which could be more optimal than the UV region.

The simulations from Section 4.14 provide a better understanding of the geometry of the nano-shells and its advantages in tuning the resonant wavelength, Moreover, Figure 16 demonstrates that as the nano-shell thickness decreases, the resonant wavelength increases and the SERS EF becomes much larger. The nano-shells geometry could be used by combining different materials such as gold and silver to receive a different resonant condition and a better chemical reaction to the solution that will be present near the nanostructures. Gold is known to be very stable in solution, whereas silver is very unstable. Further research should be done to determine the best combination of materials in the nano-shells geometry.

6.3. SERS/TERS Nano-Particles Separation

The simulations from Section 4.15 provide further insight into the effect of the mutual influence of multiple nanostructures. In the SERS/TERS simulation, a square pixel geometry was used with four hemi-spheroid nano-particles with eccentricity of $e = 0.866$ as shown in Figure 19. Figure 20 presents the mutual influence of the four nano-particle effects on the SERS EF. The separation between the nano-particles affects the SERS EF, therefore an optimal separation between the nano-particles was researched. In Figures 20–24, it appears that the separation of 110 nm between the nano-particles is the most optimal for getting the highest SERS EF. More research is required in order to determine how many nano-particles there should be per pixel and to determine the size of each pixel. Moreover, the hexagon pixel geometry should be researched in order to determine the optimal pixel shape (square or hexagon).

7. Conclusions

In this article, several directions were investigated with the final purpose of a full-scale production of a Tip-Enhanced Raman Scattering (TERS) device. Spatial distribution of enhanced electric field around metal tip for TERS was reported. The investigations focused on the effect of shape (hemi-sphere, hemispheroid, ellipsoidal cavity, ellipsoidal rod, nano-cone), and the effect of material (Ag, Au, Al) on enhancement, as well as the effect of excitation wavelengths on the electric field. The background of theoretical physics with its implementation in the simulations, yields a successful conclusion to the geometries that were analyzed. From the results section, it appears that the recommendation is for hemi-spheroid geometry for the nano-particles, and its eccentricity will be a significant parameter in the characterization of the next generation of TERS devices towards production feasibility. When analyzing the material options, silver is recommended. The use of nano-shells is a viable option for tuning the resonant wavelength of the device. To fully characterize a TERS structure, research should be directed toward combining different kinds of nano-particles geometries and their arrangement in the SERS array, in square or hexagon geometry, as previously started [44]. SERS array in hexagon geometry should be examined as well, in order to determine which geometry (square or hexagon) is better suited for enhanced performance. Additionally, optimization for separation of nano-particles and density of particles in each pixel must be performed in order to make the device's SERS EF in optimal conditions. For the next generation of TERS imagers, a beyond E^4 approximation approach must be examined in order to simulate a near-field Raman effect dipole emitter in the nano-structure vicinity. By examination of the E^4 approximation and of the extinction cross-section in various geometries, the device was accurately modeled analytically and numerically.

8. Patents

This research is the basis for several future patents.

Author Contributions: Research structure conceptualization, A.K.; investigation, Y.M. and R.M.; methodology, A.K.; project administration, A.K.; Comsol software, R.M.; supervision, Z.Z., D.Z. and A.K.; visualization, R.M.; writing—original draft, R.M. and A.K.; writing—review and editing, A.K. All authors have read and agreed to the published version of the manuscript.

Funding: This research received no external funding.

Conflicts of Interest: The authors declare no conflict of interest.

References

1. Etchegoin, P.; LeRu, E. *Principles of Surface Enhanced Raman Spectroscopy*; Elsevier: Amsterdam, The Netherlands, 2009.
2. Opilik, L.; Schmid, T.; Zenobi, R. Modern Raman Imaging: Vibrational Spectroscopy on the Micrometer and Nanometer Scales. *Ann. Rev. Anal. Chem.* **2013**, *6*, 379–398. [CrossRef] [PubMed]
3. Machida, H.; Sugahara, T.; Hirasawa, I. Preparation of dispersed metal nanoparticles in the aqueous solution of metal carboxylate and the tetra-n-butylammonium carboxylate. *J. Cryst. Growth* **2019**, *514*, 14–20. [CrossRef]
4. Fleischmann, M.; Hendra, P.J.; McQuillan, A.J. Raman Spectra of Pyridine Adsorbed at a Silver Electrode. *Chem. Phys. Lett.* **1974**, *26*, 163–166. [CrossRef]
5. Jeanmaire, D.L.; Van Duyne, R.P. Surface Raman Electrochemistry Part I. Heterocyclic, Aromatic and Aliphatic Amines Adsorbed on the Anodized Silver Electrode. *J. Electroanal. Chem.* **1977**, *84*, 1–20. [CrossRef]
6. Camden, J.; Dieringer, J.; Wang, Y.; Masiello, D.; Marks, L.; Schatz, G.; Van Duyne, R. Probing the Structure of Single-Molecule Surface-Enhanced Raman Scattering Hot Spots. *J. Am. Chem. Soc.* **2008**, *130*, 12616–12617. [CrossRef]
7. Grant, A.M.; Creighton, J.A. Anomalously Intense Raman Spectra of Pyridine at a Silver Electrode. *J. Am. Chem. Soc.* **1977**, *99*, 5215–5217.
8. Tsuneda, T.; Iwasa, T.; Taketsugu, T. Roles of silver nanoclusters in surface-enhanced Raman spectroscopy. *J. Chem. Phys.* **2019**, *151*, 094102. [CrossRef]
9. Lombardi, J.; Birke, R. A Unified Approach to Surface-Enhanced Raman Spectroscopy. *J. Phys. Chem. C* **2008**, *112*, 5605–5617. [CrossRef]
10. Ritchie, R.H. Plasma Losses by Fast Electrons in Thin Films. *Phys. Rev.* **1957**, *106*, 874–881. [CrossRef]
11. Mock, J.J.; Barbic, M.; Smith, D.R.; Schultz, D.A.; Schultz, S. Shape effects in plasmon resonance of individual colloidal silver nanoparticles. *J. Chem. Phys.* **2002**, *116*, 6755–6759. [CrossRef]
12. Witlicki, E.H.; Johnsen, C.; Hansen, S.W.; Silverstein, D.W.; Bottomley, V.J.; Jeppesen, J.O.; Wong, E.W.; Jensen, L.; Flood, A.H. Molecular Logic Gates Using Surface-Enhanced Raman-Scattered Light. *J. Am. Chem. Soc.* **2011**, *133*, 7288–7291. [CrossRef] [PubMed]
13. Lin, H.; Mock, J.; Smith, D.; Gao, T.; Sailor, M.J. Surface-Enhanced Raman Scattering from Silver-Plated Porous Silicon. *J. Phys. Chem. B* **2004**, *108*, 11654–11659. [CrossRef]
14. Talian, I.; Mogensen, K.B.; Oriňák, A.; Kaniansky, D.; Hübner, J. Surface-enhanced Raman spectroscopy on novel black silicon-based nanostructured surfaces. *J. Raman Spectrosc.* **2009**, *40*, 982–986. [CrossRef]
15. Kanipe, K.N.; Chidester, P.P.F.; Stucky, G.D.; Moskovits, M. Large Format Surface-Enhanced Raman Spectroscopy Substrate Optimized for Enhancement and Uniformity. *ACS NANO* **2016**, *10*, 7566–7571. [CrossRef] [PubMed]

16. Lu, H.; Zhang, H.; Yu, X.; Zeng, S.; Yong, K.-T.; Ho, H.-P. Seed-mediated Plasmon-driven Regrowth of Silver Nanodecahedrons (NDs). *Plasmonics* **2011**, *7*, 167–173. [CrossRef]
17. Aroca, R. *Surface-Enhanced Vibrational Spectroscopy*; John Wiley & Sons: Hoboken, NJ, USA, 2006.
18. Bao, L.-L.; Mahurin, S.M.; Liang, C.-D.; Dai, S. Study of silver films over silica beads as a surface-enhanced Raman scattering (SERS) substrate for detection of benzoic acid. *J. Raman Spectrosc.* **2003**, *34*, 394–398. [CrossRef]
19. Moskovits, M. *Surface-Enhanced Raman Spectroscopy: A Brief Perspective in Surface-Enhanced Raman Scattering—Physics and Applications*; Springer: Berlin/Heidelberg, Germany, 2006; pp. 1–18.
20. Ayas, S. Label-Free Nanometer-Resolution Imaging of Biological Architectures through Surface Enhanced Raman Scattering. *Sci. Rep.* **2013**, *3*, 2624. [CrossRef]
21. Mosier-Boss, P.A. Review of SERS Substrates for Chemical Sensing. *Nanomaterials* **2017**, *7*, 142. [CrossRef]
22. Haynes, C.L.; McFarland, A.D.; Van Duyne, R.P. Surface Enhanced Raman Spectroscopy (SERS). *Anal. Chem.* **2005**, *77*, 338A–346A. [CrossRef]
23. McNay, G.; Eustace, D.; Smith, W.E.; Faulds, K.; Graham, D. Surface-Enhanced Raman Scattering (SERS) and Surface-Enhanced Resonance Raman Scattering (SERRS): A Review of Applications. *Appl. Spectrosc.* **2011**, 825–837. [CrossRef]
24. Sharma, B.; Frontiera, R.R.; Henry, A.-I.; Ringe, E.; Van Duyne, R.P. SERS: Materials, applications, and the future. *Mater. Today* **2012**, *15*, 16–25. [CrossRef]
25. Langer, J.; Jimenez de Aberasturi, D.; Aizpurua, J.; Alvarez-Puebla, R.A.; Auguié, B.; Baumberg, J.J.; Bazan, G.C.; Bell, S.E.; Boisen, A.; Brolo, A.G.; et al. Present and Future of Surface-Enhanced Raman Scattering. *ACS Nano* **2020**, *14*, 28–117. [CrossRef] [PubMed]
26. Sonntag, M.D.; Pozzi, E.A.; Jiang, N.; Hersam, M.C.; Van Duyne, R.P. Recent Advances in Tip-Enhanced Raman Spectroscopy. *J. Phys. Chem. Lett.* **2014**, *5*, 3125–3130. [CrossRef] [PubMed]
27. Stöckle, R.M.; Suh, Y.D.; Deckert, V.; Zenobi, R. Nanoscale chemical analysis by tip-enhanced Raman spectroscopy. *Chem. Phys. Lett.* **2000**, *318*, 131–136. [CrossRef]
28. Anderson, M.S. Locally enhanced Raman spectroscopy with an atomic force microscope (AFM-TERS). *Appl. Phys. Lett.* **2000**, *76*, 3130. [CrossRef]
29. Hou, J.G.; Yang, J.L.; Luo, Y.; Aizpurua, J.; Liao, Y.; Zhang, L.; Chen, L.G.; Zhang, C.; Jiang, S. Chemical mapping of a single molecule by plasmon-enhanced Raman scattering. *Nature* **2013**, *498*, 82–86.
30. Lee, J.; Tallarida, N.; Chen, X.; Liu, P.; Jensen, L.; Apkarian, V.A. Tip-Enhanced Raman Spectromicroscopy of Co(II)-Tetraphenylporphyrin on Au(111): Toward the Chemists' Microscope. *ACS NANO* **2017**, *11*, 11466–11474. [CrossRef]
31. Tallarida, N.; Lee, J.; Apkarian, V.A. Tip-Enhanced Raman Spectromicroscopy on the Angstrom Scale: Bare and CO-Terminated Ag Tips. *ACS NANO* **2017**, *11*, 11393–11401. [CrossRef]
32. Lee, J.; Tallarida, N.; Chen, X.; Jensen, L.; Apkarian, V.A. Microscopy with a single-molecule scanning electrometer. *Sci. Adv.* **2018**, *4*, eaat5472. [CrossRef]
33. Lee, J.; Crampton, K.T.; Tallarida, N.; Apkarian, V.A. Visualizing vibrational normal modes of a single molecule with atomically confined light. *Nature* **2019**, *568*, 78–82. [CrossRef]
34. He, Z.; Han, Z.; Kizer, M.; Linhardt, R.J.; Wang, X.; Sinyukov, A.M.; Wang, J.; Deckert, V.; Sokolov, A.V. Tip-Enhanced Raman Imaging of Single-Stranded DNA with Single Base Resolution. *J. Am. Chem. Soc.* **2019**, *141*, 753–757. [CrossRef] [PubMed]
35. Crampton, K.T.; Lee, J.; Apkarian, V.A. Ion-Selective, Atom-Resolved Imaging of a 2D Cu2N Insulator: Field and Current Driven Tip-Enhanced Raman Spectromicroscopy Using a Molecule-Terminated Tip. *ACS NANO* **2019**, *13*, 6363–6371. [CrossRef] [PubMed]
36. Pettinger, B.; Domke, K.F.; Zhang, D.; Picardi, G.; Schuster, R. Tip-enhanced Raman scattering: Influence of the tip-surface geometry on optical resonance and enhancement. *Surf. Sci.* **2009**, *603*, 1335–1341. [CrossRef]
37. Thomas, S.; Wachter, G.; Lemell, C.; Burgdörfer, J.; Hommelhoff, P. Large optical field enhancement for nanotips with large opening Angles. *N. J. Phys.* **2015**, *17*, 063010. [CrossRef]
38. Karsenty, A.; Mandelbaum, Y. Computer Algebra Challenges in Nanotechnology: Accurate Modeling of Nanoscale Electro-optic Devices Using Finite Elements Method. *Math. Comp. Sci.* **2019**, *13*, 117–130. [CrossRef]
39. Jain, P.K.; El-Sayed, M.A. Plasmonic coupling in noble metal nanostructures. *Chem. Phys. Lett.* **2010**, *487*, 153–164. [CrossRef]
40. Kessentini, S.; Barchiesi, D.; Grosges, T.; De La Chapelle, M.L. Selective and collaborative optimization methods for plasmonic: A comparison. *Piers Online* **2011**, *7*, 291–295.
41. Jackson, J.D. *Classical Electrodynamics*, 3rd ed.; Wiley: New York, NY, USA, 1998.
42. Bohren, C.; Huffmann, D. *Absorption and Scattering of Light by Small Particles*; John Wiley: New York, NY, USA, 1998; pp. 357–381.
43. Halas, N. Playing with Plasmons: Tuning the Optical Resonant Properties of Metallic Nanoshells. *Mrs Bull.* **2005**, *30*, 362–367. [CrossRef]
44. Mandelbaum, Y.; Mottes, R.; Zalevsky, Z.; Zitoun, D.; Karsenty, A. Design of Surface Enhanced Raman Scattering (SERS) Nanosensor Array. *Sensors* **2020**, *20*, 5123. [CrossRef]

Article

Role of 2-^{13}C Isotopic Glyphosate Adsorption on Silver Nanoparticles Based on Ninhydrin Reaction: A Study Based on Surface—Enhanced Raman Spectroscopy

Meng-Lei Xu [1,2], Yu Gao [3], Jing Jin [2], Jin-Feng Xiong [4], Xiao Xia Han [2,*] and Bing Zhao [2,*]

1 College of Food Science and Engineering, Jilin University, Changchun 130062, China; xuml13@mails.jlu.edu.cn
2 State Key Laboratory of Supramolecular Structure and Materials, Jilin University, Changchun 130012, China; jinjing16@mails.jlu.edu.cn
3 College of Plant Protection, Jilin Agricultural University, Changchun 130118, China; gaoy1101@163.com
4 Changchun Institute of Biological Products, Changchun 130012, China; xjfzhuang@163.com
* Correspondence: hanxiaoxia@jlu.edu.cn (X.X.H.); zhaobing@jlu.edu.cn (B.Z.)

Received: 25 November 2020; Accepted: 12 December 2020; Published: 17 December 2020

Abstract: Glyphosate is one of the most commonly used and non-selective herbicides in agriculture, which may directly pollute the environment and threaten human health. A simple and effective approach to its detection is thus quite necessary. Surface-enhanced Raman scattering (SERS) spectroscopy was shown to be a very effective method to approach the problem. However, sensitivity in SERS experiments is quite low, caused by different orientation/conformation of the adsorbed molecules on the metal surface, which limit its detection by using SERS. In this paper, 2-^{13}C-glyphosate (hereafter: 13–GLP) was designed as a model molecule for theoretical and experimental studies of the molecule structure. Vibrational modes were assigned based on the modeling results obtained at the B3LYP/6-311++G** level by density functional theory (DFT) calculations, which were performed to predict the FT-IR and Raman spectra. Band downshifts were caused by ^{13}C atom isotopic substitution with mass changed. Moreover, SERS spectra of 13–GLP by combining ninhydrin reaction on Ag NPs were obtained. Isotopic Raman shifts are helpful in identifying the components of each Raman band through vibrations across the molecular system. They are coupled by probe molecules and thus bind to the substrates, indirectly offering the opportunity to promote interactions with Ag NPs and reduce the complex equilibrium between different orientation/conformation of glyphosate molecules on the metal surface.

Keywords: glyphosate; isotopic Raman shifts; SERS; ninhydrin reaction

1. Introduction

Glyphosate ($C_3H_8NO_5P$, CAS: 1071-83-6) is one of the most commonly used and non-selective herbicides in agriculture, which may directly pollute the environment and threaten human health. Glyphosate was first synthesized by the Monsanto company and marketed under the name "Roundup", and then became well known as a common herbicide to control weeds. Based on its chemical properties, glyphosate is expected to be adsorbed by the soil; due to its widespread use, it directly pollutes the environment and contaminate foods. The detection of glyphosate residues water, food, and feed stuff is an essential step in regulating and monitoring the level. The majority of these methods are based on either gas chromatography (GC) or high-performance liquid chromatograph (HPLC) coupled to a variety of detectors, mainly mass spectrometric detectors. Most of the methods are highly sensitive, however, in most cases tedious since they include numerous steps for the purification and derivatization of the compound.

Surface-enhanced Raman scattering (SERS) has been widely used as a powerful tool for ultrasensitive chemical analysis, as this technique is sensitive enough to detect glyphosate molecule [1]. Taking glyphosate as an example, although numerous efforts have been made, disagreements are still unresolved regarding the vibration assignments, and a complex equilibrium between different orientation/conformation of glyphosate molecules on the metal surface further prevents an explicit understanding of the glyphosate-detection process at a molecular level by SERS. We reported a simple and sensitive method for the determination of glyphosate by combining ninhydrin reaction and SERS spectroscopy [2]. The product (purple color dye, PD) of the ninhydrin reaction is found to SERS-active and directly correlate with the glyphosate concentration. However, vibration assignments and absorption were not completely clear.

The SERS detection method based on isotope labeling can achieve the purpose of defining vibration mode precisely. Isotope labeling has been reported in the detection of glyphosate residues and its transformation products based on NMR, MS, HPLC, GC, and GC-MS [3,4]. The SERS detection method based on isotope labeling has its own advantages. The vibration frequency of different isotopes is different, and there will be great differences. In terms of the attribution of characteristic peaks, the molecular vibration of isotope labeling was reported partly, however, the theoretical and practical spectral differences still need to be compared with DFT calculation with clearly defined assignments [5].

In this paper, in order to make more clearly defined assignments to each atom, we determined 2-^{13}C-glyphosate (13–GLP) as a model molecule for the theoretical and experimental studies of the molecule structure. Isotopic effects on Raman/FT–IR vibrational frequency and intensity due to the change of reduced masses and the vibrational coupling were verified by density functional theory (DFT) simulations and its corresponding isotopic SERS measurements. Density functional theory (DFT) calculations were performed to predict the Fourier Transform Infrared (FT–IR) and Raman spectra for the molecule. In order to reduce the complex equilibrium between different orientation/conformation of glyphosate molecules on the metal surface, a glyphosate detection method coupled by probe molecules with accurate vibration assignments thus bind to the substrates indirectly offered the opportunity to promote interactions with Ag NPs.

2. Materials and Methods

2.1. Materials

2-^{13}C-glyphosate (99%, CAS: 287399-31-9, hereafter: 13–GLP), glyphosate (99%, CAS: 1071-83-6, hereafter: 12–GLP), potassium bromide (KBr, powder, 99.999%, CAS: 7758-02-3), and silver nitrate ($AgNO_3$, 99.8%, CAS: 7761-88-8) were purchased from Sigma-Aldrich Chemical Co. (Shanghai) (Shanghai, China). Ninhydrin (2,2–dihydroxyindane–1,3–dione, $C_9H_6O_4$, CAS: 485-47-2, 98%) were obtained from Shanghai Aladdin Bio-Chem Technology Co., LTD (Shanghai, China). Sodium citrate ($Na_3C_6H_5O_7 \cdot 2H_2O$, CAS: 6132-04-3, 99%), sodium molybdate (Na_2MoO_4, CAS: 7631-95-0, 99%), and all other chemicals were analytical-grade reagents and were purchased from Beijing Chemical Reagent Factory (Beijing, China) and used without further purification. Ultrapure water (18.25 MΩ) was used throughout the experiments.

2.2. Instruments

Raman spectra were recorded on a Jobin Yvon/HORIBA LabRam ARAMIS Raman spectrometer (Tokyo, Japan) equipped with an integral BX 41 confocal microscope (Tokyo, Japan). Radiation from an air-cooled internal HeNe laser (632.8 nm) and an external cavity diode laser (785 nm) were used as the excitation source. The FT-IR spectra of ninhydrin were recorded as KBr disks at room temperature by a Bruker IFS-66V FT-IR spectrometer (Karlsruhe, Germany), equipped with a DTGS detector (Karlsruhe, Germany) at a resolution of 4 cm^{-1}.

2.3. Preparation of Silver Nanoparticles

Ag NPs were prepared according to the Lee and Meisel method, starting from silver nitrate (36 mg $AgNO_3$ added into 200 mL H_2O) and using sodium citrate (1% ω/V, 4 mL) as reducing agent [6]. After heating for 40 min at 85 °C, a grey-green colloid formed and was naturally cooled at room temperature.

2.4. Preparation of Solution

Na_2MoO_4 solution (5%, ω/V) was prepared by dissolving 5.00 g Na_2MoO_4 in 100–mL H_2O. **Ninhydrin working reagent**: ninhydrin solution (5%, ω/V) + water + acetate buffer (0.4 mol·L^{-1}, pH 5.5) (2:1:1, V/V/V). Tap water was collected without further clean-up for real sample determination.

2.5. Preparation of Purple Color Dye (PD) Product

Ninhydrin working reagent + 5% Na_2MoO_4 + 12–GLP/13–GLP solution (1:1:1, V/V). The mixed working solutions were heated in boiling water for 30 min; 10.00 μL of the producing solutions were mixed with 10 μL Ag colloid for SERS measurements [2].

2.6. Theoretical Method

All geometries were optimized using the B3LYP exchange-correlation functional, which is a hybrid of Becke's three-parameter exchange, and the Lee-Yang-Parr correlation functionals [7–9]. The 6-311++G(d,p) basis set was used for the H, C, O, N atoms. All calculations were carried out using the Gaussian 09 software program (Wallingford, CT, USA) [10]. The molecular electrostatic potential (MEP) was obtained by the Gaussian 09 software program (Wallingford, CT, USA) [11].

3. Results and Discussion

3.1. Characterization of the Ag Nanoparticles (NPs)

Ag NPs are most commonly used SERS-active substrates in fundamental and applied sciences. The spheroidal Ag NPs used in this study have a maximum absorption around 430 nm with an average diameter of 60 nm, which are consistent with the results reported in the literature (Figure S1) [12].

3.2. Molecular Geometry

The optimized geometry of 13–GLP or 12–GLP is shown in Figure 1a. In Figure 1a, ^{13}C at C7 position. The corresponding structural parameters of bond lengths, bond angles, and dihedral angles are shown in Table 1. The atom numerical labels in the following discussion refer to Figure 1a. The MEP mapping of the 13–GLP or 12–GLP is presented in Figure 1b. MEP mapping provides a visual method to understand the relative polarity of a molecule. Notably, it is a very useful procedure to study the relationship between molecular structures and their physiochemical properties. In Figure 1b, negative charges (electrophilic regions) are represented in red, while positive charges (nucleophilic regions) are represented in green; the MEP increases in the order of red < orange < yellow < green < blue. The total electron density and MEP on the complex surfaces were generated at the B3LYP/6 311++G(d,p) level of theory. 2-^{13}C-glyphosate and glyphosate are only one isotopic atom apart, thus their structural parameters are exactly the same, as shown in Table S1 (Supplementary Materials). As we know, there is no difference in space configuration between isotope labeled molecules and protomolecules, and results in this report consistent with the results reported in the literature.

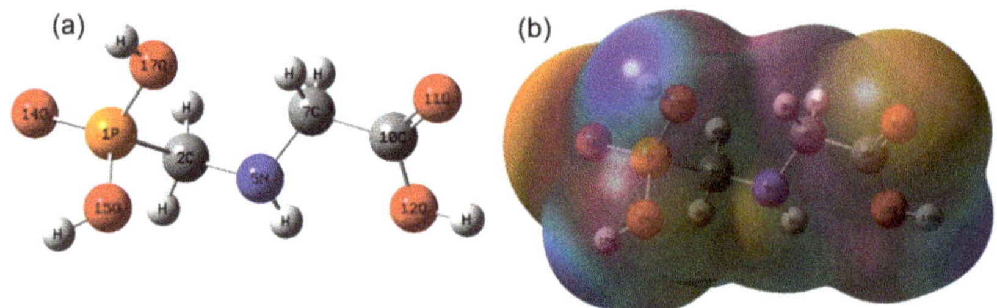

Figure 1. Optimized geometry and molecular electrostatic potential (MEP) mapping of glyphosate with/without 2-^{13}C-isotope labeled (**a**) Optimized geometry of glyphosate with/without 2-^{13}C-isotope labeled, (**b**) MEP mapping of glyphosate with/without 2-^{13}C-isotope labeled.

Table 1. Summary calculated and observed normal Raman scattering/Fourier Transform Infrared vibration modes of glyphosate (12–GLP) and 2-^{13}C-glyphosate (13–GLP).

Theoretical Vibrations Frequency (After Scaling)		Experimental Raman (Solid State)		Experimental IR (Solid State)		Vibrational Assignments
12–GLP	13–GLP	12–GLP	13–GLP	12–GLP	13–GLP	
532	531	511	504	501	501	δ(O12H) + δ(C7H_2)
629	629	648	639	648	642	τ(C2N5H) + τ(C7N5H)
674	673	-	-	-	-	ω(C10O12H)
829	829	864	857	864	856	νs(PO) + ν(PC)
856	849	918	917	916	916	C2N5C7C10 skel.
890	886	933	928	-	-	ρ(C7H_2)
979	979	979	-	982	980	ρ(C2H_2)
989	979.5	993	987	1001	997	ω(O15H) + ω(O17H)
996	995	-	1026	1032	1026	δ(HOP)
1036	1027	1037	1036	1082	1067	νs(C2N6C7)
1137	1130	1082	1069	1095	1094	νas(C2N5C7)
1310	1309	1340	1338	1335	1331	δ(C10O12H) + ω(C7H_2)
1328	1327	1422	1425	1421	1420	ω(C7H_2)
1362	1350	1432	1432	1433	1433	ω(C2H_2) + ν(C10C7)
1426	1426	1465	1460	1470	1462	δ(C7H_2)
1464	1460	1482	1482	1485	1483	ω(NH_2) + δ(C2H_2)
2799	2792	-	-	2409	2411	ν(C7H8)
3001	2992	3001	2991	3001	2922	ν(C7H9)

s: symmetry; as: asymmetry; ν: stretching vibration; δ: bending; ρ: rocking; ω: wagging; τ for twisting; skel: skeleton.

3.3. DFT Calculations of 13–GLP and 12–GLP

Each 13–GLP or 12–GLP molecule consists of 18 atoms, undergoes 48 normal modes of vibrations. The calculated vibration modes are listed in Table 1 and Table S2 (Supplementary Materials). Optimized geometry of 13–GLP and 12–GLP are the same, however, the vibrations calculated are different in some bonds.

The absence of imaginary frequencies in the DFT calculations indicates that the calculated 13–GLP and 12–GLP structures are minima on their respective potential energy surfaces. To fit the theoretical to the experimental wavenumbers, a scaling factor of 0.980 was employed to optimize the computed data under 1800 cm^{-1}, and 0.9621 for wavenumbers higher than 1800 cm^{-1}. The theoretical Raman spectra of the 13–GLP and 12–GLP calculated (scaled) at the B3LYP/6-311++G(d,p) level of theory is presented in Figure 2A(a,b), while the experimental Raman spectra in the solid state is presented in Figure 2B(a,b).

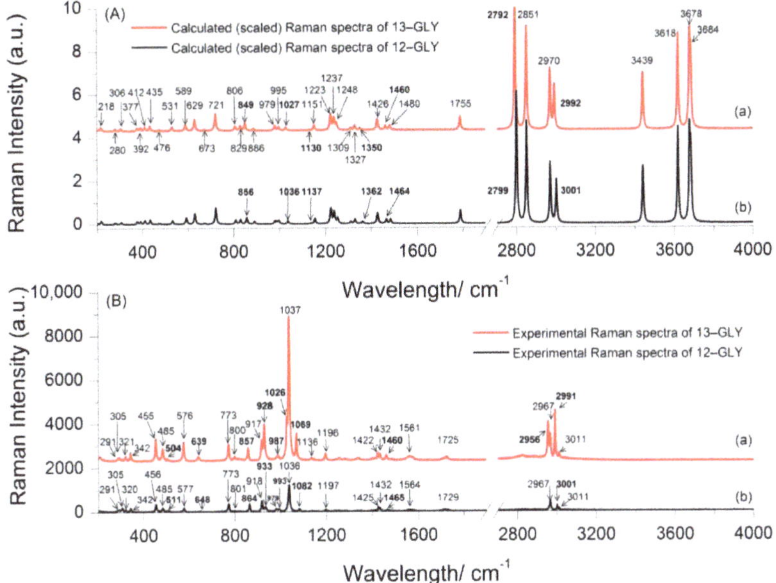

Figure 2. Experimental and calculated (scaled) Raman spectra of 2-^{13}C-glyphosate (13–GLP) and glyphosate (12–GLP) in the spectral region 200 and 4000 cm^{-1} (**A**) calculated (scaled) Raman spectra of 13–GLP (a), calculated (scaled) Raman spectra of 12–GLP (b), (**B**) Experimental Raman spectra of 13–GLP (a), Experimental Raman spectra of 12–GLP (b).

The calculated Raman modes of 13–GLP are observed at 412, 435, 476, 531, 589, 629, 673, 721, 806, 829, 849, 886, 978.6, 979.5, 995, 1027, 1130, 1151, 1223, 1237, 1248, 1309, 1327, 1350, 1426, 1460, 1480, 1755, 2792, 2851, 2970, 2992, 3439, 3618, 3678 and 3684 cm^{-1} (Table S2), Table 1 and Figure 2A(a). Results of the calculated frequencies of 12–GLP (Figure 2A(b)) were consistent with the results reported by Holanda et al. [13]. The frequencies of 13–GLP (Figure 2A(a)) in the DFT calculations indicate that most of the vibrational modes in calculated frequencies of 13–GLP agreed well with the theoretically predicted frequencies of 12–GLP. However, the isotope labeled atom changed the Raman intensity of some bonds and changed frequencies. Raman bands at 849, 1027, 1130, 1350 cm^{-1} for theoretical 13–GLP Raman spectrum appear at 856, 1036, 1137, 1362 cm^{-1} for theoretical 12–GLP assigned to C2N5C7C10 skeleton, C2N5C7 symmetric stretching, C2N5C7 asymmetric stretching, C7H$_2$ wagging + C10C7 stretching, respectively. It is noteworthy that Raman band at 1426 and 1460 cm^{-1} of 13–GLP appear at 1426 and 1464 cm^{-1} of 12–GLP assigned to $\delta(CH_2)$ from $\delta(C7H_2)$ or $\delta(C2H_2)$. Large shifts are also observed at wavenumbers higher than 1800 cm^{-1}, Raman bands at 2792, 2992 cm^{-1} for theoretical 13–GLP Raman spectrum appear at 2799, 3001 cm^{-1} for theoretical 12–GLP assigned to C7H8 stretching, C7H9 stretching. However, C7H$_2$ twisting has the same Raman spectrum appear at 1237 cm^{-1} in both 13–GLP and 12–GLP. As we know, in similar structures, the smaller the atomic mass of the chemical bond has a higher infrared absorption frequency. The isotopic substitution is expected to cause downshifts for those bands deriving from modes which involve the ^{13}C atom. Thus, a downshift for band involves ^{13}C atom isotopic substitution.

3.4. Experimental Raman Spectra of 13–GLP and 12–GLP

The observed vibration modes are listed in Table S2 (Supplementary Materials), with a tentative assignment, which is mainly based on previous work on glyphosate and related molecules [13–16]. Results of the frequencies of 12–GLP and 13–GLP (Figure 2B(a,b)) were consistent with the results reported [5]. The normal Raman modes of 13–GLP in the solid state are observed at 206, 291, 305, 320, 342, 456, 485, 511, 577, 639, 773, 800, 857, 917, 927, 987, 1026, 1036, 1069, 1136, 1197, 1255, 1283, 1338, 1425, 1432, 1460, 1482, 1564, 1729, 2956, 2967, 2991, and 3011 cm^{-1}. Most of vibrational modes observed in the solid-state Raman spectrum of 13–GLP agreed well with frequencies of 12–GLP. Summary observed normal Raman scattering vibration modes of glyphosate 12–GLP and 13–GLP are listed in Table 1. However, the downshifts for some bands cause by ^{13}C atom isotopic substitution can be assigned more accurately. As with the results of theoretical vibration, some Raman bands downshift were caused by the isotopic substitution. Raman bands at 504, 639, 928, 1069, 1350, and 1460 cm^{-1} for 13–GLP Raman spectrum in solid state appear at 511, 648, 933, 1082, and 1465 cm^{-1} for 12–GLP assigned to O12H bending and $C7H_2$ bending, C2N5H twisting and C7N5H twisting, and $C7H_2$ rocking, $C7H_2$ bending. Calculated theoretical Raman spectrum of 12–GLP/13–GLP is different from experimental Raman spectrum, the relative intensities are not precisely predicted for all the Raman bands. However, it could show structural parameters of 12–GLP/13–GLP, as they are still fairly useful for the assignments of the normal modes in the Raman spectrum. Raman spectrum will be influenced by excitation wavelength, and this may explain these discrepancies. The shapes of theoretical spectrum can well conform to experiment, but the wavenumbers are underestimated by about 15 cm^{-1}.

It is a remarkable fact that Raman shift of CH_2 bending downshift from 1465 cm^{-1} for 12–GLP to 1460 cm^{-1} for 13–GLP, another Raman shift is still 1482 cm^{-1}, however, in the calculated frequencies of CH_2 bending, one Raman shift is 1426, another downshift from 1464 cm^{-1} to 1460 cm^{-1}.

Isotopic substitution (an atom is replaced by an isotope of larger mass) leads to a decrease in the wavenumber of all modes that involve the movements of the ^{13}C atoms (harmonic oscillator: a reduces mass, μ, increases and vibrational energy level). According to the baseline of vibrational spectrum $\overline{\nu} = \frac{1}{2\pi c}\sqrt{\frac{k}{\mu}}$ shows that the larger the mass, the smaller the wavenumber. Therefore, isotopic substitution has opened the door to studying more complex systems. Thus, we assigned 1460 cm^{-1} for 13–GLP in normal Raman spectrum in solid state to $C7H_2$ bending, and 1482 cm^{-1} to $C2H_2$ bending. Similarly, we assigned 928 cm^{-1} for 13–GLP in normal Raman spectrum in solid state to $C7H_2$ rocking.

3.5. Experimental FT–IR Spectra of 13–GLP and 12–GLP

FT–IR spectrum and Raman spectrum can play complementary roles in vibration assignment. The observed vibration modes are listed in Table 1 and Table S2 (Supplementary Materials), with a tentative assignment. The normal IR modes of 13–GLP in the solid state are observed at 473, 501, 579, 642, 781 and 798, 829, 856, 916, 980, 997, 1026, 1067, 1094, 1167, 1202, 1223 and 1244, 1269, 1331, 1420, 1433, 1462, 1483, 1560, 1709 and 1730, 2411, 2536, 2833, 2922, 2991 cm^{-1}. Most of vibrational modes observed in the solid-state IR spectrum of 13–GLP agreed well with frequencies of 12–GLP. However, the downshifts for some bands cause by ^{13}C atom isotopic substitution can also be observed (Figure 3). IR bands at 642, 997, 1331, and 1462 cm^{-1} for 13–GLP in solid state appear at 648, 1001, 1335, and 1470 cm^{-1} for 12–GLP assigned to C2N5H twisting + C7N5H twisting, O15H wagging + O17H wagging, C10O12H bending+ $C7H_2$ wagging, $C7H_2$ bending. The downshifts for these bands cause by ^{13}C atom isotopic substitution with mass changed.

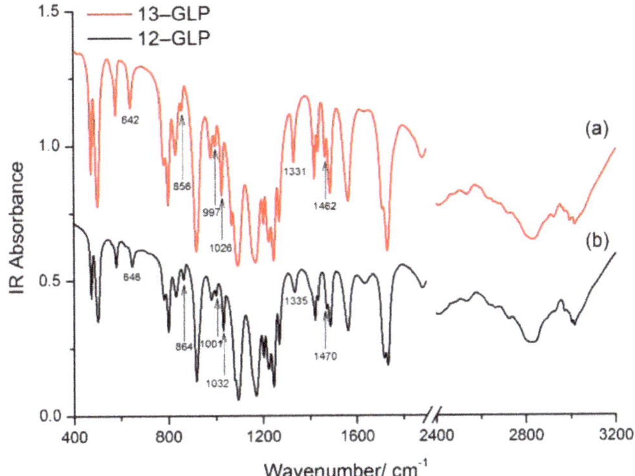

Figure 3. Experimental Fourier transform infrared (FT–IR) spectra of 2-^{13}C-glyphosate (13–GLP) and glyphosate (12–GLP) in the spectral region 400 and 4000 cm^{-1} (**a**) IR spectra of 13–GLP, (**b**) FT–IR spectra of 12–GLP.

The band at 864/857 cm^{-1} assigned to PO symmetric stretching and PC stretching also exist in Raman/IR shift with no bond to C7 atom. Feis et al. [5] consider (CO_2)–(CNH_3) as two-mass oscillator, downshift cause by mass changed. Likewise, the IR band at 1032/1026 cm^{-1} assigned to HOP bending also downshifted, yet only Raman shift of 13–GLP at 1026 cm^{-1} is observed, Raman shift of 12–GLP is not observed.

In conclusion, spectra shift by isotopic can be assigned vibrational modes more accurately because it does not change the symmetry of molecular structure and force constants between chemical bonds. The vibrational frequency of isotope substitution is shifted due to the difference of mass. In this paper, bands downshifts cause by ^{13}C atom isotopic substitution to ^{12}C.

3.6. Raman and SERS Spectra of PD POroduct from 12–GLP/13–GLP

Glyphosate Raman and SERS spectra have been subject of different experimental and theoretical studies, we also obtained SERS spectra of 13–GLP absorbed on Ag NPs. However, the observed SERS pattern in terms of a complex equilibrium between different orientation/conformation of the glyphosate molecules on the metal surface shows that sensitivity in SERS experiments is quite low [5].

Pesticide molecules without SERS-active functional groups could be coupled by some probe molecules to achieve a better signal [17]. In previous work, we reported a glyphosate detection method based on ninhydrin reaction and surface–enhanced Raman scattering spectroscopy (SERS) to overcome this disadvantage. The PD product of the ninhydrin reaction is found to SERS-active and directly correlate with the glyphosate concentration. Meanwhile, experimental information about the structure changes of glyphosate and its PD product absorbed on Ag nanoparticles still needs further detection. Normal Raman and SERS spectra of PD product from 12–GLP/13–GLP with vibrational assignments are shown in Figure 4 and Table S3. They are clearly distinguishable with two relatively stronger peaks around 662 and 791 cm^{-1} in each spectroscopy. Figure 4a,c compares the Raman spectra of PD product from 13–GLP and 12–GLP are very similar, yet two band at 662 and 791 cm^{-1} display intensity changed due to mass changes caused by the ^{13}C atom isotopic substitution.

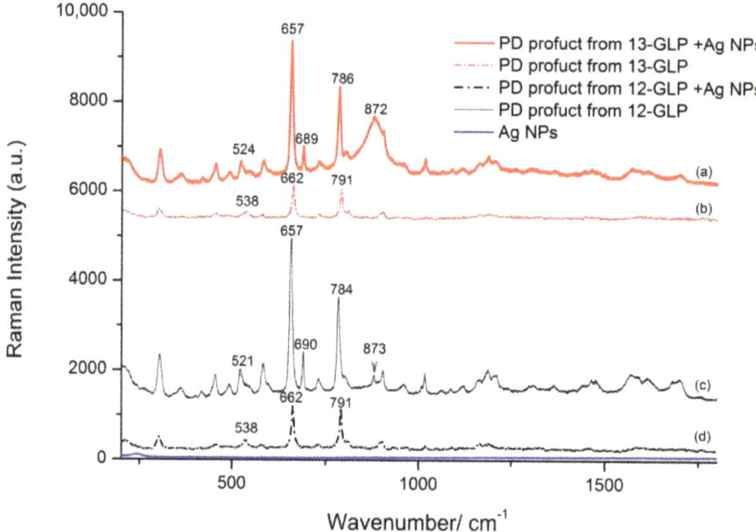

Figure 4. Raman and Surface-enhanced Raman scattering (SERS) spectra absorbed onto Ag NPs of the purple color dye (PD) products from glyphosate (12–GLP) and 2-^{13}C-glyphosate (13–GLP) at a concentration of 1.0×10^{-3} mol·L^{-1}. (**a**) SERS spectra of PD products from 12–GLP (**b**) Raman spectra of PD products from 12–GLP (**c**) SERS spectra of PD products from 13–GLP (**d**) Raman spectra of PD products from 13–GLP.

Furthermore, SERS spectra of PD product from 12–GLP/13–GLP are different when absorbed on Ag NPs. PD product from 12–GLP/13–GLP absorbed on Ag NPs, there are slight shifts of the two strong bands from 662 and 791 cm^{-1} in the normal Raman spectra to 657 and 784/786 cm^{-1}, respectively. Raman intensity of the band at 662 and 791 cm^{-1} of normal PD product from 12–GLP/13–GLP are similar, however, SERS intensity are different, the peak at 662 cm^{-1} is much higher than at 784/786 cm^{-1}. The band at 538 cm^{-1} shift to 521/524 cm^{-1} when absorbed on Ag NPs in SERS spectra. A strong peak appears at 690/689 cm^{-1} in SERS spectra that is not present in experimental Raman spectra (Figure 4). The peak at 872/873 cm^{-1} assigned to Mo–O from Na_2MoO_4 which is a catalyst forms a weak π-complex with aromatic rings of the ninhydrin component in the PD product. This peak in PD product from 13–GLP is much higher with wider full width of half maximum than from 12–GLP.

This reaction is different from the classic amino acid color reaction with ninhydrin, in which no decarboxylation or dealdehyding reaction takes place. The molecular structure of PD product includes two moieties, N5 atom from glyphoate combines with ninhydrin, forming a new C−N bond and linking glyphoate and ninhydrin. Ninhydrin molecular is C_2 symmetry, however, in PD product changed to C_1 symmetry. The peaks at 657 and 786/784 cm^{-1} are assigned to the out–of–plane bending vibration of C=O from ninhydrin molecular. The peak 689 and 524/521 cm^{-1} are related to the vibration of the benzene ring carbon skeleton. This suggests that PD product interacts with the Ag atoms via the oxygen atoms of the ninhydrin moiety. According to the surface selection rule, we infer the PD product molecules adsorbed on the surface had a perpendicular orientation. ^{13}C atom isotopic substitution to ^{12}C change the mass, thus PD product coupled MoO_4^{2-} with a weak π-complex slightly changes the adsorption orientation on Ag NPs.

Glyphosate Raman and SERS spectra have been the subject of different experimental and theoretical studies but sensitivity in SERS experiments is quite low due to small Raman cross-section and complex equilibrium between different orientation/conformation of the adsorbed molecules on the metal surface. –COO– group does not participate in the interaction of the glyphosate with Ag NPs. Imidazole ring in PD product interacts with the Ag NPs surface, adopting almost erect state orientation relative to it. None of these bands can be attributed to the vibrations of glyphosate, therefore these moieties do not affect the SERS spectrum of glyphosate on the surface of Ag NPs. Thus, the combination of high Raman scattering probe with isotopic labeling provides avenues for accurate and quantitative SERS. We believe such a method will open a new promising area for isotopic SERS to investigate the mechanism of surface photoinduced reactions through tracking the fingerprint information changes of the species on metal nanostructures.

3.7. *Quantitative Analysis of 13–GLP*

We examined the SERS spectra of the PD products from 13–GLP at different concentrations shown in Figure 5. The SERS intensity at 657 cm^{-1} shows a good linear relation with the concentrations of PD product in the range of 5.0×10^{-7}–1.0×10^{-4} mol·L^{-1} ($y = 6535.9 + 1027.8 \times \lg C$, $R^2 = 0.9487$). The limit of detection (LOD) is thus calculated to be 4.3×10^{-7} mol·L^{-1}. Limit of quantitation (LOQ) is 5.0×10^{-7} mol·L^{-1}. Recoveries for tap water (1.0×10^{-6} mol·L^{-1}, 5.0×10^{-5} mol·L^{-1}, 1.0×10^{-5} mol·L^{-1}) were 83.8–115.5%, relative standard deviation (RSD) were 1.1-16.8%.

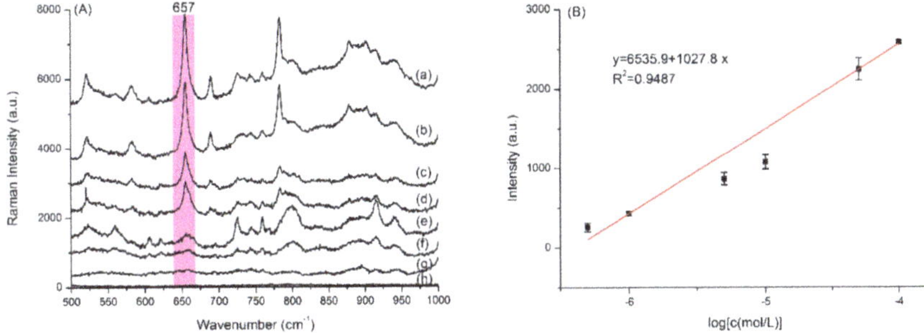

Figure 5. (**A**) Representative concentration-depend SERS spectra of (a) 1.0×10^{-4}, (b) 5.0×10^{-5}, (c) 1.0×10^{-5}, (d) 5.0×10^{-6}, (e) 1.0×10^{-6} mol·L^{-1}, (f) 5.0×10^{-7} mol·L^{-1}, (g) 1.0×10^{-7} mol·L^{-1} purple color dye product from 2-^{13}C-glyphosate (13–GLP) and (h) blank; the standard curve (**B**). Excitation wavelength: 633 nm.

4. Conclusions

Glyphosate Raman and SERS spectra have been the subject of different experimental and theoretical studies, however, sensitivity in SERS experiments is still quite low due to the complex equilibrium between different orientation/conformation of the adsorbed molecules on the metal surface. 2-^{13}C isotopically substituted derivative is used to confirm the vibration modes in Raman and FT-IR spectra. The isotopic substitution is expected to cause downshifts deriving from modes which involve the ^{13}C atom, and PO symmetric stretching and PC stretching with no bond to ^{13}C atom. This downshift caused by isotopic substitution is also present in DFT calculations, though there is no difference in space configuration between isotope labeled molecules and protomolecules. We further reported the SERS spectra of 13-GLP by combining ninhydrin reaction on Ag NPs coupled by probe molecules and thus bind to the substrates. This indirectly offers the opportunity to promote interactions with Ag NPs and reduce the complex equilibrium between different orientation/conformation of glyphosate molecules on the metal surface with a LOD of 4.38×10^{-7} mol·L^{-1} and LOQ of 5.0×10^{-7} mol·L^{-1}. 2-^{13}C isotopically substituted derivative is used to confirm the vibration modes in Raman and FT-IR spectra more conveniently with higher sensitivity, better reproducibility, and lower relative standard deviation. This work will open a new promising area for isotopic SERS to investigate the mechanism of surface photoinduced reactions through tracking the fingerprint information changes of the species on metal nanostructures.

Supplementary Materials: The following are available online at http://www.mdpi.com/2079-4991/10/12/2539/s1, Figure S1: UV-Vis spectrum of the Ag NPs, Table S1: The optimized structural parameters of 2-13C-glyphosate or glyphosate calculated at the B3LYP/6-311+G** level, Table S2: Calculated and normal Raman spectra of glyphosate (12-GLY) and 2-13C-glyphosate (13-GLY) in the frequencies, Table S3: Raman and SERS band assignments (cm−1) of the Purple Color Dye (PD) product from glyphosate (12–GLP) and 2-13C-glyphosate (13–GLP).

Author Contributions: Methodology, M.-L.X. and Y.G.; Software, M.-L.X., J.J. and J.-F.X.; Writing—Original Draft Preparation, M.-L.X. and X.X.H.; Writing—Review and Editing, M.-L.X. and B.Z.; Project Administration, M.-L.X., X.X.H. and B.Z. All authors have read and agreed to the published version of the manuscript.

Funding: This research was funded by National Natural Science Foundation of People's Republic of China, grant number 22011540378, 21773080, 21711540292, the Development Program of the Science and Technology of Jilin Province, grant number 20190701003GH), the China Postdoctoral Science Foundation, grant number 2018M631876, the Open Research Fund of National Research Center of Engineering and Technology of Tea Quality and Safety, grant number 2017NTQS0201. The APC was funded by National Natural Science Foundation of People's Republic of China 22011540378.

Conflicts of Interest: The authors declare no conflict of interest.

References

1. Tu, Q.; Yang, T.; Qu, Y.; Gao, S.; Zhang, Z.; Zhang, Q.; Wang, Y.; Wang, J.; He, L. In situ colorimetric detection of glyphosate on plant tissues using cysteamine-modified gold nanoparticles. *Analyst* **2019**, *144*, 2017–2025. [CrossRef] [PubMed]
2. Xu, M.L.; Gao, Y.; Li, Y.; Li, X.; Zhang, H.; Han, X.X.; Zhao, B.; Su, L. Indirect glyphosate detection based on ninhydrin reaction and surface-enhanced Raman scattering spectroscopy. *Spectrochim. Acta A* **2018**, *195*, 78–82. [CrossRef]
3. Markus, A.M.; Kraus, M.; Miltner, A.; Hamer, U.; Novak, K.M. Effect of temperature, pH and total organic carbon variations on microbial turnover of $^{13}C_3{}^{15}N$-glyphosate in agricultural soil. *Sci. Total Environ.* **2019**, *658*, 697–707.
4. Chan, Y.; Fei, W.; Sae, J.C.; Jun, Y.; Ruth, E.B. Phosphate oxygen isotope evidence for methylphosphonate sources of methane and dissolved inorganic phosphate. *Sci. Total Environ.* **2018**, *644*, 747–753.
5. Feis, A.; Gelini, C.; Ricci, M.; Tognaccini, L.; Becucci, M.; Smulevich, G. Surface-enhanced Raman scattering of glyphosate on dispersed silver nanoparticles: A reinterpretation based on model molecules. *Vib. Spectrosc.* **2020**, *108*, 103061. [CrossRef]
6. Lee, P.C.; Meisel, D. Adsorption and surface–enhanced Raman of dyes on silver and gold sols. *J. Phys. Chem.* **1982**, *86*, 3391–3395. [CrossRef]
7. Becke, A. Density–functional exchange–energy approximation with correct asymptotic behavior. *Phys. Rev. A* **1988**, *38*, 3098–3100. [CrossRef]
8. Becke, A. Density–functional thermochemistry. III. The role of exact exchange. *J. Chem. Phys.* **1993**, *98*, 5648–5652. [CrossRef]
9. Lee, C.; Yang, W.; Parr, R. Development of the Colle–Salvetti correlation–energy formula into a functional of the electron density. *Phys. Rev. B* **1988**, *37*, 785–789. [CrossRef] [PubMed]
10. Frisch, M.J.; Trucks, G.W.; Schlegel, H.B.; Scuseria, G.E.; Robb, M.A.; Cheeseman, J.R.; Scalmani, G.; Barone, V.; Mennucci, B.; Petersson, G.A.; et al. *Gaussian 09 (Revision A.02)*; Gaussian, Inc.: Wallingford, CT, USA, 2009.
11. Bulat, F.A.; Toro-Labbe, A.; Brinck, T.; Murray, J.S.; Politzer, P. Quantitative analysis of molecular surfaces: Areas, volumes, electrostatic potentials and average local ionization energies. *J. Mol. Model.* **2010**, *16*, 1679–1691. [CrossRef]
12. Gao, Y.; Xu, M.-L.; Xiong, J. Raman and SERS spectra of thiamethoxam and the Ag3–thiamethoxam complex: An experimental and theoretical investigation. *J. Environ. Sci. Health B* **2019**, *54*, 665–675. [CrossRef]
13. Rocicler, O.; Holanda, C.B.S.; Daniel, L.M.V.; Paulo, T.C.F. High pressure Raman spectra and DFT calculation of glyphosate. *Spectrochim. Acta. A* **2020**, *242*, 118745–118752.
14. Yael, J.A.; Fuhr, J.D.; Bocan, G.A.; Millone, A.D.; Tognalli, N.; Afonso, M.S.; Martiarena, M.L. Abiotic Degradation of Glyphosate into Aminomethylphosphonic Acid in the Presence of Metals. *J. Agric. Food Chem.* **2014**, *62*, 9651–9656. [CrossRef]
15. Parameswari, A.; Asath, R.M.; Premkumar, R.; Benial, A.M.F. SERS and quantum chemical studies on N–methylglycine molecule on silver nanoparticles. *J. Mol. Struct.* **2017**, *1138*, 102–109. [CrossRef]

16. Costa, J.C.S.; Ando, R.A.; Sant'Ana, A.C.; Corio, P. Surface–enhanced Raman spectroscopy studies of organophosphorous model molecules and pesticides. *Phys. Chem. Chem. Phys.* **2012**, *14*, 15645–15651. [CrossRef] [PubMed]
17. Xu, M.-L.; Gao, Y.; Han, X.X.; Zhao, B. Detection of Pesticide Residues in Food Using Surface-Enhanced Raman Spectroscopy: A Review. *J. Agric. Food Chem.* **2017**, *65*, 6719–6726. [CrossRef] [PubMed]

Publisher's Note: MDPI stays neutral with regard to jurisdictional claims in published maps and institutional affiliations.